The Greenwich Guides to Astronomy

The *Greenwich Guides* are a series of books on astronomy for the beginner. Each volume stands on its own but together they provide a complete introduction to the night sky, everything it contains, and how astronomers are discovering its secrets. Written by experts from the Old Royal Observatory at Greenwich, they are right up to date with the latest information from space exploration and research and are suitable for observers in both the northern and southern hemisphere.

Available now
The Greenwich Guide to Stargazing
The Greenwich Guide to The Planets
The Greenwich Guide to Stars, Galaxies and Nebulae
The Greenwich Guide to Astronomy in Action

The Old Royal Observatory at Greenwich, London, is open daily to visitors. It is the home of Greenwich Mean Time and of the Greenwich Meridian which divides east from west. It also houses the largest refracting telescope in Great Britain.

For more information write to: The Marketing Department, National Maritime Museum, Greenwich, London SE10 9NF.

The Greenwich Guide to

Astronomy in Action

Carole Stott

George Philip
in association with
The National Maritime Museum, London

British Library Cataloguing in Publication Data

Stott, Carole
    The Greenwich guide to astronomy in action.
    1. Astronomy
    I. Title
    520

ISBN 0-540-01184-3

© The Trustees of The National Maritime Museum 1989
First published by George Philip Ltd,
59 Grosvenor Street, London W1X 9DA

Printed in Hong Kong

Acknowledgements

I am grateful to many people for their assistance in the preparation of this book. I would like to thank Marilla Fletcher, Stuart Malin, Harry Ford and other colleagues at the Old Royal Observatory, Greenwich. Special thanks go to my husband, David Hughes of Sheffield University, and Lydia Greeves of George Philip, who have both been exceptionally supportive.

I am also grateful to Paul Doherty for producing the illustrations on pp. 11, 18 (top and bottom), 19 (bottom), 20 (top and bottom), 24, 25 and to the following for permission to reproduce their illustrations: The Anglo-Australian Telescope Board p. 31 (top and bottom); Professor Roy L. Bishop, Acadia University, Canada p. 19 (top); Lawrence Englesberg, Jodrell Bank p. 33; The European Southern Observatory pp. 12–13, 26, 29, 40, 43; The European Space Agency p. 56; Istituo e Museo di Storia della Scienza, Firenze p. 17; Max-Planck-Institut für Radioastronomie p. 8; NASA/JPL pp. 15 (bottom), 45, 57, 60, 61, 62, 63 (left and right), 65, 66, 69, 70, 71, 73, 74, 74–5, 78–9, 80, 81, 83, 84 (left and right), 85, 86, 87, 88, 89, 91; The Trustees, The National Maritime Museum, London pp. 36, 37, 38, 39; National Optical Astronomy Observatories pp. 2, 9 (top and bottom), 15 (top), 42, 46, 47, 48, 49 (bottom); Royal Greenwich Observatory, Herstmonceux p. 23 (David Calvert); Royal Observatory, Edinburgh pp. 41, 49 (top), 50, 55, 67; Science Photo Library (SPL), London pp. 30, 32, 34–5, 53; Jim Stevenson pp. 16, 21, 54; X-ray Astronomy Group, University of Leicester p. 59.

Jacket illustrations: NASA/JPL (front); Jim Stevenson (back).

FRONTISPIECE *Most of the world's leading observatories are built on high-altitude sites scattered round the globe. The instrumentation they contain includes some 150 large telescopes with diameters of at least 1 metre. At the Kitt Peak National Observatory in Arizona shown here, the largest telescope is the 4-metre Mayall, housed in the dome on the right. The other domes contain 36-inch and 90-inch telescopes.*

JACKET ILLUSTRATIONS *An artist's impression of IRAS, the infra-red satellite launched in 1983 (front), and the Old Royal Observatory, Greenwich, London (back).*

Contents

Introduction

Astronomers are at the forefront of Man's quest to understand our fascinating and ever-changing Universe, using the most sophisticated and up-to-date techniques to collect and analyse information from the cosmos and to interpret the wide range of bodies it contains, from our sister planets to the distant and enigmatic quasars. This book introduces those techniques and what we have learned from them, showing how observations with the powerful successors to Galileo's simple telescope of 1609 are used in combination with the startling revelations of the X-ray, infra-red and ultra-violet Universe to advance our understanding of our environment. The continual struggle to improve observing conditions is also explored, from new observatories placed high in remote mountain areas to the efforts to send instruments into space and Man's dream of travelling among the planets himself. If you have ever wondered what an astronomer does, or wanted to be one yourself, here is a complete introduction to this mind-stretching world.

1 · An Astronomer's Life

There have always been astronomers. The very first humans regarded the heavens with awe and curiosity and monuments such as Britain's Stonehenge suggest that early man endeavoured to understand his celestial environment. These first astronomers used the positions of the heavenly bodies to mark the course of the day and record the changing seasons, and in time they were used to guide travellers. Three hundred years ago when the Royal Observatory at Greenwich was in its infancy astronomers were still primarily concerned with where the heavenly bodies were, rather than with their nature, but the emphasis of their studies gradually changed. By the mid nineteenth century the objects themselves had become the focus of interest – what they were made of, where they had come from and how they would develop and change. These are still the major questions today. Although many of the puzzles that astronomers wrestled with in the past have now been answered, these questions have been replaced by new ones as larger telescopes are developed, new techniques employed, and new discoveries made. Astronomy is a dynamic and exciting subject and there is no wonder that many people want to get involved and play their part.

But what does an astronomer do? How are current problems being tackled? Do you have an image of a grey-haired gentleman spending the night hours twiddling the controls of a large telescope while he peers intently through the eyepiece at some twinkling star, pausing only occasionally to curse the impending clouds and sip his coffee, trying to keep awake. Well, you would be wrong. Present-day astronomers include both men and women. They use some of the most sophisticated equipment ever devised and are often immersed among the largest and most powerful computers. Nine-tenths of astronomers round the world are on the staff of universities, teaching as well as doing research. The other 10 per cent work in government establishments. What may seem extraordinary is that most of them spend very little time actually observing – it is possible to collect enough observational information from an object in, say, a week or two to keep an astronomer busy for a year. On the other hand it is essential that the information astronomers are working on is as good as possible. They travel the world to find the best observing sites, and the telescopes they use, whether on Earth or in space, are triumphs of engineering.

The amount of astronomical information available is now so vast that no astronomer could possibly study everything. Instead, most professionals specialize in an area of interest, falling into one of a number of recognized groups. The three principal categories of astronomer are those concerned with the stars and interplanetary dust, those interested in galaxies, and those involved with the planets, whose field includes the Earth beneath our feet, as well as comets, asteroids and interplanetary space. A further category includes those concerned with theories, such as the science of celestial mechanics, centred on the movements of the heavenly bodies, or cosmology, which is concerned with explaining the origin, evolution and future of the Universe as a whole.

Gone are the days when an astronomer would spend night after night peering through the eyepiece of his telescope making continual adjustments to improve the view. Large observatory telescopes, such as the 100-metre radio instrument at Effelsberg in Germany seen here, are computer controlled; the astronomer sits in comfort at the control console.

The cosmologist constructs a theoretical model of the Universe which can then be checked against observational work carried out by other astronomers. The two theoretical models that have been investigated this century are the more popular Big Bang theory (which postulates that the Universe started in the explosion of a compact ball of matter and has been expanding ever since) and the steady state theory (which postulates that the Universe had no beginning and will never end, but will stay 'steady' in space and time, with matter being continuously created to fill the gaps formed by expansion). The steady state theory suggests that the Universe thousands of millions of years ago appeared to be the same as it is now, whilst the Big Bang theory envisages a more compact Universe in the past with the stars and galaxies much closer together than they are now. The observations of distant objects support the Big Bang theory, so this is the one that most cosmologists now favour and continue to work on. Cosmologists should not be confused with cosmogonists, who are specifically concerned with the origin of solar systems, rather than with the development of the Universe as a whole, although the only known solar system is our own. Two other important groups are made up of specialists in instrumentation and those who popularize astronomy.

ABOVE *Astronomers can be found in the most unexpected places. This group are erecting a solar telescope in the Antarctic to take advantage of the continuous hours of sunlight in the summer months.*

So today most astronomers are specialists, spending the majority of their time on one small area of astronomical interest. On the other hand they all need to know about other aspects of astronomy and how their particular specialization fits in. It is impossible to work successfully in isolation. If, for example, you were interested in rock on Mars, you would not get very far by looking at the planet through a telescope. But

Modern optical observatories are sited to get the best view possible. The Kitt Peak Observatory shown here is high up in the Quinlan Mountains of Arizona.

9

you would, hopefully, soon be able to take advantage of material brought back to Earth from the Martian space probes. Your conclusions on the Martian rock would be of interest to scientists studying other planets, meteorites, the formation of the Solar System, and those looking for planetary systems round stars other than our Sun. Or you could be interested in objects which are much further away from our Earthly base, such as a particular type of star or a grouping of stars, a cluster or a galaxy too distant to be visited by our spacecraft. For these we have to rely on information gleaned from what we can see from the vicinity of Earth, either from our planet itself, or with the aid of instruments in orbit around Earth. The astronomer may not need to do any observing himself, as observations made by other astronomers at the world's leading observatories are often accessible to those interested. The modern astronomer needs to keep abreast of his colleagues' work and also to publish his own work so that others can benefit.

It is also sometimes possible to make use of observations made in the past. One of the most exciting astronomical spectacles is a supernova, a star at the end of its life cycle which suddenly flares up in a great explosion. But there have only been three supernovae in our Galaxy in the previous millennium, all of which occurred before the telescope had been invented, which meant that astronomers interested in these phenomena worked with naked-eye observations made in 1054, 1572 and 1604 by Chinese and European astronomers. Naturally enough, when another supernova burst forth unexpectedly in 1987 in the Large Magellanic Cloud, a companion galaxy to our own, all available telescopes and instrumentation were trained on it, for nobody knows exactly when another such explosion will occur.

Information from astronomical objects comes in a variety of forms, most of which are kinds of electromagnetic radiation. The best known form of radiation and the one we are all most familiar with is light, but objects in space also give off radio waves, microwaves, infra-red radiation, ultra-violet waves, X-rays and gamma rays, all of which can tell us something about the objects concerned. All electromagnetic radiation travels at the same speed – the speed of light, 299,792 kilometres per second – and is most usually envisaged as moving in an undulating wave-like way. The distance between the crests of the waves is known as the wavelength, and it is this which distinguishes one type of electromagnetic radiation from another. Light, for example, has a wavelength of a few hundred nanometres (a nanometre = 10^{-9}, or .000,000,001 metres), whereas the wavelength of radio waves is measured in terms of metres. Differences in the wavelength of light itself give us the whole spectrum of colours in visible light, ranging from a wavelength of 400 nanometres at the blue end of the spectrum to just over 700 nanometres at the red end. The whole range is seen in a rainbow, where the colours are arranged in order of decreasing wavelength: red, orange, yellow, green, blue, indigo, violet.

Only some of these different types of information actually reach the astronomer on Earth. Generally speaking, short wavelength radiation – ultra-violet radiation, X-rays and gamma rays – cannot penetrate the Earth's atmosphere. But even some of the light waves from a star can be absorbed by the atmosphere, and it affects all types of radiation to some extent. The only way to eliminate this interference is to place astronomical instruments above the Earth's obscuring gases. As the diagram shows, a whole range of instruments, many of which do their work far above us, has been developed to investigate the various kinds of radiation.

Specialized instrumentation and techniques have been developed to investigate the full range of information reaching us from objects in the heavens. Some are Earth-based, such as optical and radio telescopes, and some are carried by balloons and satellites far above us.

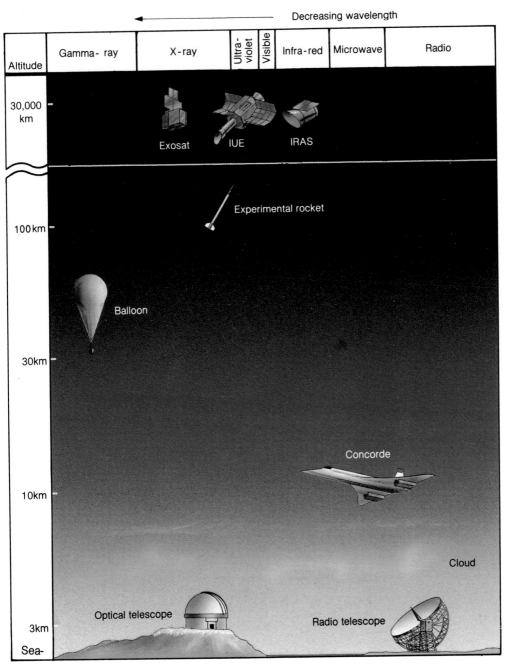

Decreasing wavelength

Altitude	Gamma-ray	X-ray	Ultra-violet	Visible	Infra-red	Microwave	Radio

30,000 km

Exosat   IUE   IRAS

Experimental rocket

100 km

Balloon

30 km

Concorde

10 km

Cloud

Optical telescope

Radio telescope

3 km

Sea-level

11

Just as there are astronomers specializing in different kinds of astronomical objects, so there are those interested in particular forms of radiation and the detection techniques involved. A radio or infra-red astronomer, for example, might work on one object for two or three years and then switch to another. And there are space astronomers who specialize in the use of astronomical techniques aboard satellites and spacecraft in the environment of space. All these different methods of investigation will be used by the object-orientated astronomers. A 'dustman' — an astronomer interested in interstellar dust — will be part theoretician, part optical observer, part space scientist, part infra-red astronomer and lots of other parts mixed up.

By using the whole range of electromagnetic radiation we are able to get a more complete picture of the Universe. We can learn much more about the objects we can see as well as investigate objects which are otherwise invisible. Those curious non-objects known as black holes come into this category, regions of space where the power of gravity is so strong that not only matter but also electromagnetic radiation within a certain distance is sucked in and cannot escape (see *The Greenwich Guide to Stars, Galaxies and Nebulae*). We cannot 'see' the black hole itself, but can detect it by observing objects accelerating as they are drawn towards it and other objects caught within its gravitational field. The occurrence of black holes has been indicated theoretically, but there is still controversy over whether they exist. We are also unable to 'see' many of the radio galaxies. Although some have a visible galaxy at their centre, the radio signals which enable us to detect these objects come from a much larger area than the light source.

Some of the most beautiful and detailed views of the sky are produced by traditional photography, such as this view of a spiral galaxy in the constellation of Sculptor. An artificial satellite was captured in the long exposure, its path shown by the red line.

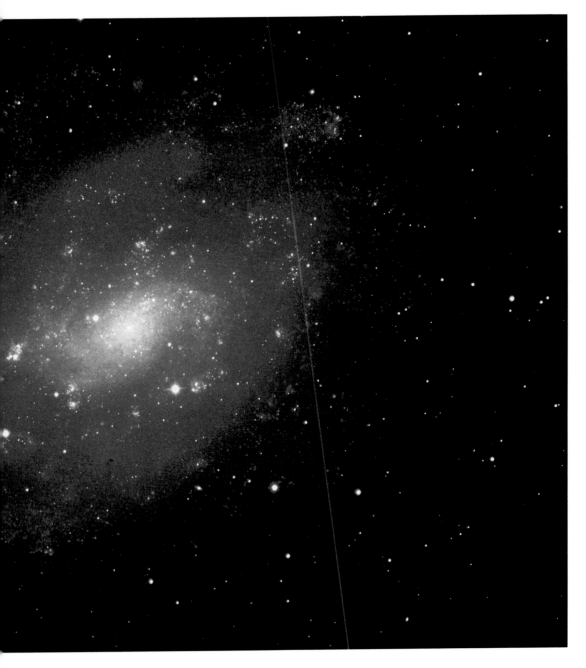

Taking all the groups discussed together, there are about 6000 professional astronomers in the world, about 500 of whom work in the United Kingdom. It might seem surprising that the United Kingdom's professional society of astronomers, the Royal Astronomical Society (RAS), currently (1988) has about 2500 Fellows, 148 of whom are women, but some of these are from countries other than the United Kingdom, whilst others would describe themselves in another way, as geophysicists or mathematicians rather than as astronomers. Many, too, are people who are actively interested in astronomy but are not professionals. Of those who earn a living from their interest in astronomy, most will be found working for either a government-backed establishment, such as the observatories at Herstmonceux and Edinburgh, the Rutherford Appleton Laboratory, Chilton, or perhaps the British Space Agency, or for one of the universities which has a relevant department, such as Cambridge, with its Institute of Astronomy, or Manchester, which includes the Nuffield Radio Astronomy Laboratories incorporating Jodrell Bank.

Each of the government organizations has built up expertise and is pursuing research in specific areas of interest, for which they are now well known. The Royal Greenwich Observatory (RGO), for example, has a long tradition of very high-quality work in positional astronomy (the branch of astronomy concerned with the positions of the heavenly bodies), so many astronomers joining the team would spend time in this field of study. The RGO is also responsible for running the United Kingdom's Northern Hemisphere Observatory (NHO) on La Palma in the Canary Islands. So an RGO astronomer would have opportunities to use these facilities for his own research as well as helping visiting astronomers to the NHO to carry out theirs. Like all other astronomers he would then endeavour to have his findings published and to present them at an astronomical conference.

The best Earth-based observing sites are scattered round the globe. This is the Cerro Tololo Observatory in the Andes of central Chile.

Once he has satisfied his teaching requirements – lectures, tutorials and research work with students who may be the astronomers of the future – an astronomer attached to a university has more freedom in the choice and execution of his area of study. It may be that there is nobody else at the university working on the same astronomical subject, or that he is one of a small group of staff working in this specific area. The radio astronomers working with the Jodrell Bank telescopes attached to Manchester University contrast with the cometary astronomers dotted about the country at Birmingham, Cardiff, Sheffield, Leicester and London universities. They are also lucky in that they have sophisticated equipment on their doorstep. Telescopes in British universities, however, are usually primarily devoted to student experiments rather than to research work, so professional astronomers attached to universities have to travel to find the high-class up-to-date facilities and ideal observing conditions they need. On the other hand they may only need to use the very costly modern telescopes and other equipment they require for just a few nights a year. Most large observatories are now visited by scientists from all over the world and the use of their facilities is very carefully planned and regulated.

By a certain date each year, every astronomer in the United Kingdom who is interested in working with one of the government-financed telescopes applies to the department that is ultimately responsible for the country's astronomical facilities, the Science and Engineering Research Council (SERC). Each application is considered by a committee which is largely made up of other United Kingdom astronomers, as specialist knowledge is essential to co-ordinate the requests and choose the most appropriate. Research applications can usually fill the time

available ten times over, but if an astronomer is lucky enough to have his project selected the council will then provide all the necessary funding, covering air tickets and subsistence as well as facilities required for the research itself. The same system would apply to those who wanted to use an instrument in orbit around the Earth, such as the International Ultra-Violet Explorer (IUE). Here the available observing time is shared between the bodies that funded the instrument concerned. In the case of the IUE, an astronomer could apply to either the European

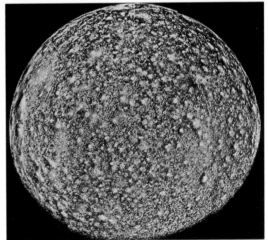

Images obtained from spacecraft, such as this close-up view of Jupiter's moon Callisto, have increased enormously our knowledge of our neighbours in space.

15

Space Agency (ESA), or to SERC, or to the National Aeronautics and Space Administration (NASA), who all have a proportion of observing time to allocate. Should an astronomer want to further his research by having an instrument or experiment carried on a space probe, it is necessary to apply to the body which is designing and launching the craft. In this way British and American astronomers are helping with the experiments carried on the Soviet *Phobos* missions to Mars and many British experiments have been carried on ESA missions.

The true professional astronomer, earning his

Amateur astronomers enjoying the night sky. Many amateurs make observations of use to professionals.

living by investigating the heavens, is far outnumbered by the amateurs. Many amateurs, however, have a great knowledge of and dedication to astronomy and collect valuable astronomical data. In Britain most of these people belong to the British Astronomical Association (BAA), although some may also be Fellows of the Royal Astronomical Society (RAS), just as many RAS Fellows are also BAA members. The BAA has a number of sections composed of people following a particular area of interest. Many of these carry out projects that are not covered by professional astronomers and that involve repeated observations with a relatively small instrument (such as a 10-inch telescope), which will be used night after night from the amateur's home provided observing conditions are good. Observers in the comet section, for example, supply regular observations of a particular comet, and it is often through them that new and previously unknown comets are discovered. The observations of these groups of dedicated amateurs are highly regarded by the professional astronomers who do not have the time to carry out this regular work.

The largest group of all are the true amateurs, people who are interested in astronomy just as a hobby and are keen to go sky-watching, or to keep up with the latest in the subject by attending lectures or watching television programmes like 'The Sky at Night', but who do not have the dedication or desire to take this interest more seriously. This group gains enormously from the efforts of another kind of professional astronomer, those who work in the media, writing articles and presenting television and radio programmes that interpret what the university and observatory astronomers are doing, and talking about astronomy in terms that the layman can understand. Or they look after visitors to astronomical centres open to the public. The best known such 'media' astronomer in the United Kingdom is Patrick Moore.

2 · Looking

The telescope remains the astronomer's most important instrument, not because every astronomer uses one but because they all rely on the results from one, whether it is Earth-based, in orbit, or aboard a space probe. It is also the most familiar of all the astronomer's tools. The instrument that everybody knows, the optical telescope, collects and magnifies the light it receives to give an observer a better view of whatever he or she is looking at. But the word 'telescope' is now used to apply not only to instruments that 'look' at the heavens, but also to those that detect non-visual messages from space. The 'listening' telescope is known as the radio telescope and will be looked at in Chapter 3.

Although the optical telescope is one of the astronomer's oldest aids, it can also be described as a relatively recent invention. Man has been interested in the heavenly bodies ever since he first gazed up at the sky and observed the planets and the stars, but the first time anyone used a telescope to get a better view was in 1609. It is not possible to say who actually invented the device, but, contrary to what every schoolboy knows, the Dutch spectacle-maker Hans Lippershey rather than Galileo is usually credited with constructing the first effective one. And it is thought the telescope was first used on the battlefield, the advantages of being able to see your enemy from a safe distance before he could see you being quickly realized.

Although Galileo did not invent the instrument with which he is always associated, he did play a dominant role in the telescope's early application to astronomy, and the observations that he made in the second half of 1609 were to change Man's views of the Universe. Of equal importance was the fact that he published his results and did so quickly, summarizing them in his *The Starry Messenger* as soon as March 1610. This ensured that news of his discoveries soon spread throughout Europe. The telescope had shown him that the Moon had mountains and lowlands like the Earth, evidence that our planet could no longer be regarded as a unique body. The Milky Way was transformed from the band of light that is seen with the naked eye into hundreds and thousands of separate stars, and the telescope also revealed that there are many, many more stars in all other areas of the sky than could be seen without the aid of an instrument.

Galileo's telescopes in the History of Science Museum in Florence. His observations established that the Solar System is centred on the Sun rather than revolving round the Earth.

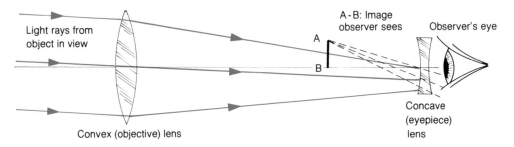

Galileo was also responsible for making the important discovery that Jupiter has satellites orbiting their parent planet. These observations were among the first to support the view that the Solar System is centred on the Sun, rather than revolving around the Earth as was generally believed at the time. Further conclusive evidence followed in 1610 when he observed that Venus has phases much like our own Moon, a phenomenon that could only be explained if the Solar System was Sun-centred. Galileo was also to see that Saturn has 'ears', but it was only towards the end of the seventeenth century that telescopes became good enough to resolve these 'ears' into the planet's now familiar rings.

Galileo and other contemporary astronomers, such as the Englishman Thomas Harriot and the German Christopher Scheiner, used a very simple telescope involving two lenses, one convex and one concave (see diagram). The convex lens in this telescope is also known as the objective lens as it gathers the light from the object being

Galileo's telescope used a convex lens to collect the light from astronomical objects such as the Moon. The light was focused by the concave eyepiece lens to produce a magnified image (A–B).

viewed. The concave or eyepiece lens then directs the light rays to a focus, producing the image that the observer sees. The magnifying power of the eyepiece lens determines how much the image being viewed will be enlarged. Galileo could magnify an image up to thirty times, the degree of enlargement that is given by some of the more powerful binoculars in use today. His telescope had the disadvantage that the area of sky he could see through the telescope at one time – the

Johannes Kepler's improved telescope design used two convex lenses, one to collect the light from the object being viewed (the objective lens) and the other to form a magnified image (the eyepiece lens). The image seen by the observer is upside down, but this is no problem for an astronomer.

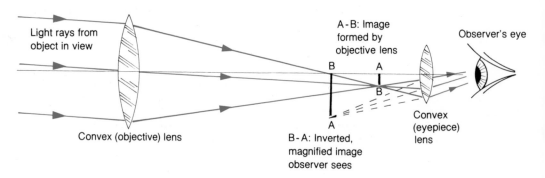

These rainbows arching over Woolsthorpe Manor in Lincolnshire where Sir Isaac Newton was born symbolize his important discovery that white light is made up of the colours of the spectrum.

field of view – was not very large, a drawback resolved by the German Johannes Kepler's improved design of 1611 (see diagram), which used two convex lenses. As the diagram shows, this telescope produces an inverted image. Although this is not a problem for astronomers, seeing everything upside down was a drawback on the battlefield, so Kepler's telescope did not have the same commercial appeal as the Galilean design, although it did produce the quality of image required by astronomers.

Another great name in telescope design is that of the Englishman Sir Isaac Newton. Following his work with white light, which showed that it is a combination of light of different colours (the colours of the rainbow mentioned in Chapter 1), he realized that when light passes through a telescope lens each colour focuses at a slightly different point (see diagram), so producing a discoloured image in the telescope eyepiece. This

distortion, known as chromatic aberration, does not occur when mirrors are used to collect and focus the light rather than lenses. The mirror (reflector) telescope which Newton developed consequently gave a clearer image than the lens (refractor) telescope. When Newton's instrument was exhibited at the Royal Society in 1672 it aroused great interest. He was not, however, the only person working on the design of the reflector, the Scottish astronomer James Gregory and the Frenchman Cassegrain both producing their own designs – Gregory in 1663, although his instrument was not built until some years later, and Cassegrain in 1672. The Newtonian and Cassegrain designs are shown on p. 20.

From the first days of the telescope in the seventeenth century many astronomers, scientists and engineers have attempted to improve on these early instruments and in many cases have been successful. They have overcome the problems of producing a lens-system to cope with chromatic aberration, and of perfecting the lenses and mirrors themselves. They have also experimented with these components in different combinations to produce telescopes with the

Sir Isaac Newton was the first to realize that the colours of the spectrum each focus at a slightly different point when light passes through the telescope lens, so producing a distorted image. This problem of chromatic aberration does not occur with mirror telescopes.

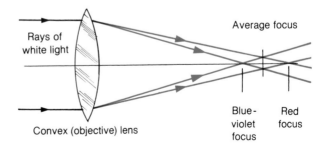

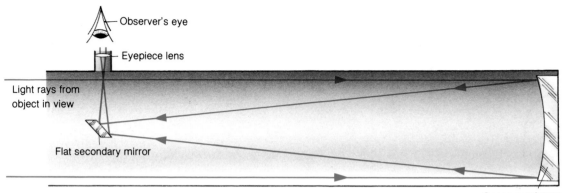

Observer's eye
Eyepiece lens
Light rays from object in view
Flat secondary mirror
Concave primary (objective) mirror

Sir Isaac Newton's mirror telescope was an improvement on the early lens instruments. A concave mirror collects the light and reflects it back down the telescope tube to a flat secondary mirror which focuses it to form an image.

finest possible images and optimum light-gathering power.

As technology has progressed and astronomical interests have changed, different types of telescope have risen and fallen in popularity. Until the nineteenth century the small refractor (with a lens of up to 10 inches in diameter) was the most popular, although there were those who preferred reflectors, most notably the discoverer of Uranus, Sir William Herschel, and the Irish astronomer, Lord Rosse, who successfully made and used large instruments of this type.

Herschel's 48-inch mirror telescope was the largest reflector in the world for several years until it was superseded by Lord Rosse's 72-inch aperture instrument in 1845. As lens-making techniques improved, larger refracting telescopes were constructed. The 28-inch refractor at Greenwich, completed in 1894, was one of a number produced in the last decade of the nineteenth century. By this date photography was being used increasingly to record observations through telescopes and many instruments were being built with a photographic plate-holder in place of the eyepiece. The

Many of the original telescope designs, like this one by the Frenchman Cassegrain, are still used today. Here a concave mirror collects the light and a convex mirror directs it to a focus.

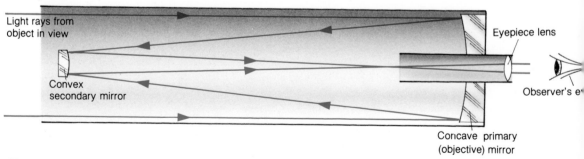

Light rays from object in view
Eyepiece lens
Convex secondary mirror
Observer's e*
Concave primary (objective) mirror

Greenwich 28-inch telescope could be used for either method of observation by altering the lens system within the telescope tube and changing the apparatus at the eyepiece end. A number of similarly large refractors were built in the United States, the biggest of which, the Yerkes Observatory refractor some 130 kilometres north-west of Chicago, was completed in 1897. With a lens of 40 inches, this is still the largest refracting telescope in the world.

Large refractors were very popular and successful around the end of the nineteenth and the beginning of the twentieth centuries. This type of instrument was used to determine the first accurate distances to the stars and to ascertain the stars' motions and chemical constituents. At the same time the high magnification given by the best eyepieces facilitated detailed work on the Moon and the planets, and gave astronomers the best Earth-based observations of these bodies. The results achieved were only surpassed when astronomers were able to send apparatus into space above the Earth's atmosphere.

With the Yerkes 40-inch telescope the refractors had reached their technical limit. As the instruments got bigger, the lenses became thicker as well as larger and absorbed increasing amounts of light from the object under observation. They also became increasingly heavy, and as they could only be supported round the edge they tended to sag in the middle. By comparison, there are no such technical difficulties in using large mirrors as these can be supported from behind. Coupled with the fact that new mirror-making techniques, such as depositing silver on glass and, more recently, aluminium on glass, give increasingly good mirror finishes, these advantages have brought reflectors back into fashion. Reflectors are also cheaper to mount than refractors because they are more compact, an important consideration when the mounting is usually as expensive as the telescope itself.

When a 100-inch reflector was constructed on Mount Wilson in southern California in the early years of this century, and proved to be successful, several more instruments were built, one of the finest being the great Mount Palomar 200-inch Hale telescope, also in southern California, completed in the late 1940s. Weight and expense were saved without any loss of quality by casting the Pyrex mirror blank as a cellular structure rather than as a solid disc. The largest reflector in the world is the Soviet 236-inch completed in 1976, although there have been a number of sizable reflectors built since then, such as the 4-metre Mayall at the Kitt Peak National Observatory in the Quinlan Mountains of Arizona.

Similar telescopes are at Cerro Tololo in Chile and at the Anglo-Australian Observatory in eastern Australia. They all use the Ritchey-Chretien system, a modification of the Cassegrain design, which allows the telescope to cover a wide area of sky with good definition. One of the newest reflectors is the 4.2-metre William Herschel telescope constructed over a number of years and first used regularly in 1987 as part of the Northern Hemisphere Observatory complex on La Palma in the Canary Islands. It is named after Sir William Herschel in recognition of his pioneering work, and has a primary mirror 4.2 metres in diameter, which makes it the third largest single mirror telescope in the world.

Although currently popular, reflectors are not without their drawbacks. Any misalignment of the mirrors will result in distortion and they are particularly sensitive to the flexure and vibration of the tube and to temperature effects. This means they are inferior to refractors in work involving high precision measurements and they generally offer a smaller field of view. But the increase in their light-gathering power has enabled astronomers to see further and further into space.

Observing and recording ever more distant objects is of major importance in attempting to understand the origin of the Universe itself, as it is these very remote bodies which enable us to look back in time to the period when the Universe was taking shape after the Big Bang (see *The Greenwich Guide to Stars, Galaxies and Nebulae*). To do this astronomers need large aperture telescopes with the necessary light-gathering power. Increasingly sensitive light detectors (see Chapter 5) have been applied to these telescopes to gather as much information as possible, but, as in the case of refractors, the upper size limit of reflectors using a single mirror has now nearly been reached. Above a certain diameter, distortion inevitably creeps in. The most recent instrument designs aim to overcome size limitations by constructing a large mirror out of a mosaic of smaller mirrors, or by having an aperture made

One of the world's finest and newest telescopes is the 4.2-metre William Herschel on the Canary Island of La Palma. It is supported and manoeuvred on an altazimuth mount.

up of several independent telescopes on the same bearings, or by combining the light from a number of totally independent telescopes. By 1980 a 180-inch multi-mirror telescope using six individual mirrors to make up the primary mirror was assembled and operating at Mount Hopkins in Arizona. Although this is not the largest reflector ever made, it points the way for future developments. Of equal importance, it was built for a cost considerably below that normally needed for a telescope of this aperture and it has been housed in an inexpensive box-shaped building which rotates with the telescope rather than in the traditional, much more costly, hemispherical dome.

By the early 1990s the Keck telescope being constructed on the summit of Mauna Kea in Hawaii should be in operation. This instrument's unique honeycomb array of 36 hexagonal mirrors is equivalent to a single mirror 10 metres across, giving four times the light-gathering capacity of the 200-inch Hale telescope.

The instrument known as the Schmidt photographic telescope, named after Bernhard Schmidt who first described it in 1932, is a combination of the reflector and the refractor, offering a wide field of view. Telescopes based on this design are used by observatories all over the world. The 48-inch Schmidt at the Palomar Observatory has been used to make one of the most detailed photographic maps of the northern sky. This involved some 2000 photographs and includes very faint stars down to about magnitude 20.

Telescope mounts

However good a telescope is at gathering light and producing a sharply focused image, it is no

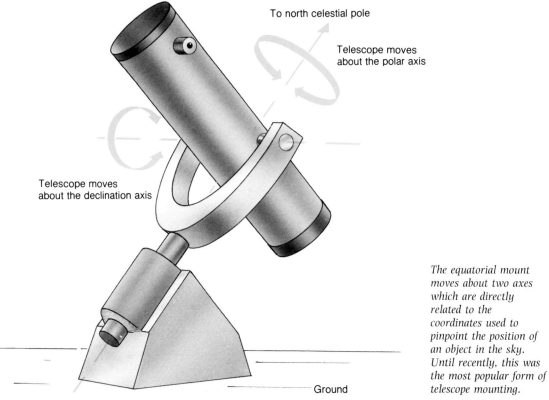

To north celestial pole

Telescope moves
about the polar axis

Telescope moves
about the declination axis

Ground

The equatorial mount moves about two axes which are directly related to the coordinates used to pinpoint the position of an object in the sky. Until recently, this was the most popular form of telescope mounting.

use at all unless it is supported by a mount which will point it accurately to the chosen position in the sky. The two systems widely in use – the equatorial and the altazimuth – both rotate about two axes so that the telescope can be pointed to any position in the heavens. The equatorial mounting rotates about what are known as the polar and declination axes (see diagram). These are directly related to the co-ordinates used to describe the position of an astronomical object in the sky, its right ascension (RA) and declination, the longitude and latitude of the celestial sphere (see *The Greenwich Guide to Stargazing*). RA is altered by rotating about the mount's polar axis, which is set up parallel to a line linking the two celestial poles. The de-

clination axis is at right-angles to the polar axis and can be adjusted according to the required declination of the object under view (see diagram). Once the equatorially-mounted telescope has been set up to point at the chosen object, it can be moved to follow the body in its apparent progress across the sky. By rotating the polar axis at the same speed but in the opposite direction to the rotation of the Earth, the object being studied will stay in the telescope's field of view.

Two of the most popular equatorial mountings are the English and the German. The English mounting has been used to carry a number of large refractors, including the 28-inch refractor at Greenwich, but it has the disadvantage that a small region of sky around the pole cannot be observed. The German mounting, on the other hand, can operate over the whole sky and has proved a very popular choice with the smaller and lighter refractors (with lenses of less than 15 inches).

The equatorial mount has been the more popular for the past 150 years, largely because it is relatively easy to use it to track a heavenly object, which is something that every astronomer needs to do. The altazimuth mount, on the other hand, has to be driven simultaneously in both axes if it is to keep an object in view, and it has the added complication that the drive rate needs to be varied throughout the night and according to the position of the object. Until recently this has discouraged astronomers from using it, although it is in fact the simplest of

The altazimuth mount moves vertically and horizontally to align the telescope.

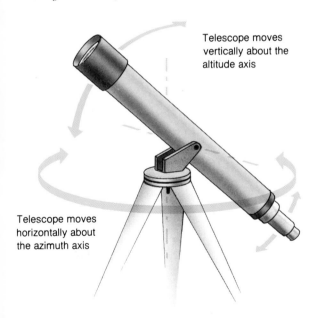

Telescope moves vertically about the altitude axis

Telescope moves horizontally about the azimuth axis

mounts, moving purely vertically and horizontally about what are known as the altitude and azimuth axes (see diagram). However, now that the drive can be regulated by computer, altazimuth mounts are becoming more popular and they are being used for many of the new large telescopes, such as the 4.2-metre William Herschel.

Altazimuths are now not only easy to use but also have the advantage that they are smaller than the equatorial mounts and involve substantial savings in the overall cost of an observatory telescope and in the construction of the observatory dome. The William Herschel telescope is operated from a computer console, where the astronomer keys in the position of the object chosen for observation. As the telescope moves automatically to point to the object, the dome rotates with it, so that the instrument's view is always clear. Once the object is located its image appears on the console screen for observation and the telescope continues to track it across the sky. Constructing such a large telescope is a very lengthy and costly business. It has been estimated that a night's observing on one of the world's finest telescopes, like the William Herschel, costs £8000 on average, so it is no wonder that governments join together when building new facilities and that requests for observing time are carefully vetted.

Your own telescope

Should you want a telescope of your own it need not cost quite so much as the William Herschel. An amateur astronomer does not need to buy the most expensive type of instrument, nor one which offers the greatest magnification. The important question is what you want to use your telescope for; are you more interested in exploring the surface of the Moon or looking at distant nebulae? Your interests and the amount of money you have available should dictate what you buy. Money pays for good optical com-

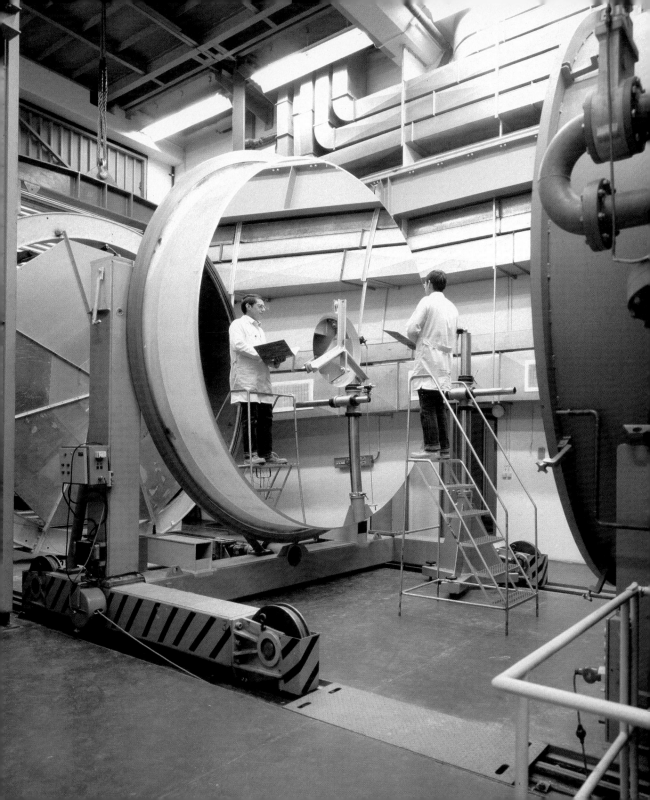

ponents, a steady mount and a telescope's general finish. Of these, the first two qualities are the important ones, as it is the performance and not the appearance of the telescope that you should be interested in. So, make sure that you have the opportunity to test how bright and sharp the image is when the telescope is focused on some familiar scene such as a distant building or tree, and that the mount is good and sturdy. If the tube containing the optics is 'bobbing' about, you will be unable to keep the object in view.

Of all a telescope's features it is the size and quality of the main light-gathering element that is most important, whether it is a lens or a mirror. The diameter of this light-gathering component is usually used to describe the whole telescope, as in a 9-inch refractor. Some early instruments, however, are known by their focal length, the distance between the main lens or mirror and the point where the image is formed. For example, the instrument with which William Herschel discovered the planet Uranus is always described as a 7-foot reflector, referring to the focal length. This measurement is also an indication of the length of the telescope tube.

An instrument's magnifying power can be changed by simply changing the eyepiece, but remember that a higher magnification that would give you a more detailed view of a lunar crater also covers a smaller area of the Moon's

The most important part of a telescope is its light-gathering component. Every few years this mirror from the 3.6-metre telescope at the European Southern Observatory in Chile is stripped of its aluminium coating and re-aluminized to ensure the instrument is in prime condition.

surface than you would see with lower magnification. A good starting-point would be a reflector with an aperture of 10 to 15 centimetres, large enough easily to show the principal satellites of Jupiter and Saturn and to give a detailed view of the lunar craters and nearby galaxies. Ever popular is the catadioptric reflector. The simple spherical surface of this telescope's main mirror can be mass-produced without any quality loss, so it represents good value for money. It also has the advantage that it is compact, easy to move from one position to another or to carry in the boot of a car. Your local astronomical society will be full of people with practical experience who can help you decide between the many different models on offer. They can also help you with information on making your own telescope if you would prefer to do this.

Small refractors and reflectors with apertures of 5 centimetres or less are to be avoided. They may seem a bargain in the shop and their claims to high magnification can be very tempting, but you would be better off buying binoculars. These are in fact a very good instrument for any astronomer to start with. They are cheaper than a telescope, more compact, more portable and equally useful for both astronomical and terrestrial viewing. Using binoculars is also an excellent way to get to know the night sky; even when you have progressed on to something larger, you will find you still use them regularly. If you do decide to purchase binoculars go for a good all-purpose pair with a magnifying power of 7 and an aperture of 40 millimetres (described as 7 × 40). Remember, too, that their performance can be greatly improved by mounting them on a camera tripod.

27

3 · Listening

As we go about our daily lives, we rely on all our senses to inform us about our environment. Eliminate one of these and the picture you get is incomplete. Just think how different your impression of a television programme is when the sound is turned down. Similarly, interpreting the Universe only on the basis of what we can see gives us a one-sided view, and our understanding has been greatly advanced by interpreting information coming to us in forms of electro-magnetic radiation other than light. These have revealed a strikingly different Universe.

Radio wavelengths were the first of these forms to be investigated by astronomers. Although the existence of radio waves has been known since 1887 through the work of Heinrich Hertz, and Thomas Edison suggested these could be picked up from the Sun as long ago as 1890, radio astronomy was not born until half a century later when Karl Jansky, working in the United States in 1932, first detected cosmic radio noise. Jansky had not been expecting to make this discovery. At the time he was investigating the background noise in radio receivers and it was while listening to the crackles generated during thunderstorms that he first heard the steady hissing sound caused by radio waves from space. He then discovered that this reached its maximum volume when his aerial was pointing towards the centre of our Galaxy where the concentration of stars is greatest. Although his work received much publicity at the time, including a radio programme which broadcast the hissing of the cosmos to the people of the United States, it took another ten years before radio astronomy really got under way. During this time only one man – Grote Reber – worked as a radio astronomer. He built the first radio telescope, a 30-foot steerable parabolic reflector, in his own backyard and recorded cosmic radio waves.

Another contributor in the earliest years of radio astronomy was J. S. Hey, who worked in a British research team during World War II. To their surprise the team found that radar equipment was being jammed, not by Germany as was first thought, but by radio waves from the Sun. This radiation came from an active sunspot, but Hey and his colleagues were also to detect radar echoes from meteor trails and in 1948 identified the radio source in the constellation of Cygnus now known as Cygnus A, still one of the most powerful radio galaxies known to Man. This early work gave birth to investigations of radio emissions from the Sun at the Sydney and Cambridge radio observatories, which in turn led to the study of many other radio sources. There are now so many sources known that it is impossible to continue the system of identifying the source by giving it the name of its parent constellation and a letter of the alphabet, as in Cygnus A, and observatories around the world now produce their own catalogues using their own identification symbols.

As the diagram on p. 11 shows, radio waves lie

This 15-metre radio telescope has been listening in to very short radio signals from space since 1987. Part of the European Southern Observatory in Chile, it is the only large submillimetre telescope in the southern hemisphere.

at one end of the electromagnetic spectrum. As everyone who listens to the radio will know, they cover a wide range of wavelengths, from waves of 30 metres and more at one end to submillimetre waves of only one-third to 1 millimetre in length at the other. The shortest wavelengths are absorbed by the Earth's lower atmosphere and the longer end of the range is reflected by the upper atmosphere, but a large range in between reaches us here on Earth and can be studied from virtually anywhere.

Today radio observatories throughout the world listen in to the signals reaching Earth from a variety of cosmic sources. The familiar radio dish telescopes work in much the same way as an optical telescope. Large, smooth metal surfaces are used to collect and reflect the radio waves, rather as mirrors gather light in the large reflectors, but the collecting area of a radio telescope needs to be much bigger than that of an optical instrument because radio wavelengths are almost a million times longer than those of visible light. To resolve the same kind of detail as an optical telescope the radio telescope needs to be roughly a million times bigger. The waves recorded are then focused to a point where an antenna transforms them into electrical signals. These are amplified to increase the strength of the signal and then analysed by a computer to build up a map of the radio source. These images look very similar to topographical maps of the Earth's surface, although the 'contours' indicate areas of equal brightness rather than equal height, as they would on a conventional map. Recently-developed techniques mean it is now also possible to display a radio source on a black and white screen to show how it would look through radio eyes, giving the illusion of a three-dimensional object. This image can then be computer-enhanced in the same way that optical pictures are treated.

Radio astronomy offers a new way of looking at apparently familiar objects and has also revealed hitherto unknown phenomena. Seen

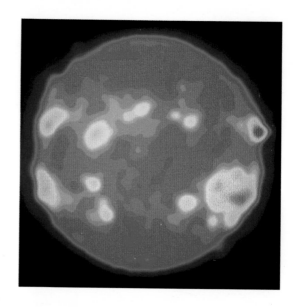

ABOVE *A radio view of the Sun obtained from the 100-metre Effelsberg radio telescope. The red active regions are clouds of hot gas which would be revealed as sunspots on an optical photograph. The weakest signals are denoted by the blue ring round the Sun.*

through a radio telescope, an elliptical galaxy suddenly increases in size. Where there was thought to be only empty space on either side, lobe-like emissions of radio waves can be detected, which may span an area of space millions of light years across. In many of these radio galaxies it is also possible to detect a small radio source at the galaxy's core. The third strongest radio source in the sky is Centaurus A, a galaxy containing a million million stars which is easily seen from Earth with a pair of binoculars.

The galaxy Centaurus A is one of the brightest as viewed from Earth. It is also a source of strong radio, X-ray, gamma-ray and infra-red radiation. Both these views were produced from the same photographic plate, but that below was 'massaged' to give greater detail of the central dust lane.

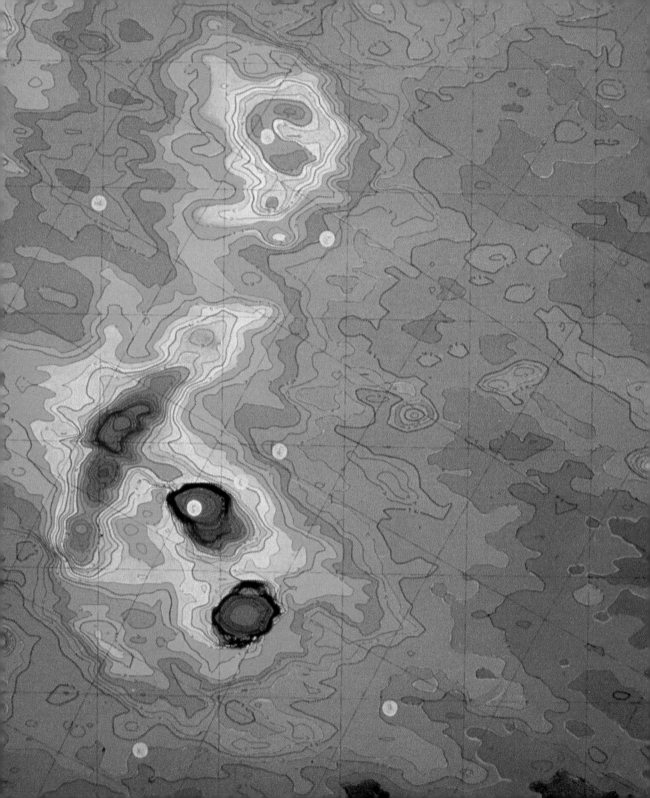

The familiar constellation of Orion looks quite different when seen through 'radio' eyes, the strong radio emission from the Horsehead and Orion Nebulae showing as two deep red areas on and below Orion's belt. The stars superimposed on this image give the familiar optical outline.

But only radio telescopes reveal the lobes stretching out from its optical centre. The brightest object in the sky at radio wavelengths is Cassiopeia A, seen only as a dimly glowing gas cloud on an optical photograph. Cassiopeia A is in fact an expanding shell of gas around the remains of a supernova, its fainter outer layers concealing an inner shell of more brightly glowing material.

Even more exciting are the new objects discovered through radio astronomy, including two of the most enigmatic and puzzling celestial phenomena yet identified – quasars and pulsars. Pulsars – the rotating remnants of massive collapsed stars – were first discovered in 1967 by Jocelyn Bell at Cambridge, who identified the short regular pulsations produced by one of these objects by combining signals from over 2000 individual radio aerials. In fact the radio signals are so regular that they were initially thought to be being transmitted by extraterrestrial beings, and were known for a short time as LGMs (Little Green Men). Since then it has been realized that these radio sources can be identified with the rapidly spinning condensed objects known as neutron stars, the spin rate being responsible for the regularity and brevity of the radio signals. The fastest yet discovered is the Crab pulsar in the centre of the Crab Nebula, first identified in 1968. It revolves every 0.033 seconds (although it is now slowing down) and has a magnetic field a million million times more powerful than that of the Earth.

The curious phenomena known as quasars (short for quasi-stellar radio source) are more puzzling. First identified in 1963 using radio observations, they appear star-like when viewed in optical wavelengths. But they are still a mystery. They are very distant, very compact objects emitting energy at prodigious rates. It is thought they may be the furthest objects we can see, thus taking us back to an early stage in the development of the Universe (see *The Greenwich Guide to Stars, Galaxies and Nebulae*).

One of the nearest quasars known is 3C 273, the twentieth strongest radio source in the sky. The exact position of 3C 273 was pinpointed in 1962 by observing exactly when signals were intercepted by the Moon passing in front of it. This piece of research also revealed that there are two closely-spaced radio sources involved.

Like optical telescopes, radio telescopes need to be able to move in order to point to the chosen area of sky. The Mark 1A telescope at Jodrell Bank, one of the world's leading radio observatories, is a fully steerable 76-metre paraboloid

The Jodrell Bank Mark 1A 76-metre dish is the second largest fully steerable radio telescope in the world.

dish built in 1957 under the guidance of Bernard Lovell. It is second only to the 100-metre diameter dish at Effelsberg in Germany, which is the largest fully steerable radio telescope and marks the upper limit in size for an instrument of this type. Telescopes larger than this, such as the Arecibo 305-metre dish in Puerto Rico, use the rotation of the Earth to sweep the heavens. The Arecibo instrument, the largest single dish in the world, is built in a natural hollow in the hills of the island and because of its size can detect some very faint radio sources. Even so, it is only capable of resolving detail equivalent to that perceived by the human eye at visible wavelengths, because radio waves are so much longer than those of the visible spectrum. By slightly changing the orientation of the telescope's receiver every day, it can be used to produce a radio map of the whole available sky.

In the same way as a mosaic or series of mirrors is used to increase the light-collecting area of an optical telescope, a number of smaller radio telescopes can be used to act as one large one. This technique is known as interferometry. Martin Ryle at Cambridge produced one of the first instruments of this type in the 1960s, composed of two fixed antennae 800 metres apart and a third which was movable along an 800-metre track. Like the Arecibo telescope, the Cambridge instrument also relies on the rotation of the Earth to survey the sky. Each antenna collects radio waves in the normal way which are then fed into a computer to reconstruct the shape of the radio source. The world's largest purpose-built interferometer is the Very Large Array (VLA), 80 kilometres west of Socorro in New Mexico, which consists of 27 movable dishes each 25 metres in diameter set out in the form of a Y. Together they produce a resolving power equivalent to an optical telescope with a diameter of 27 kilometres, allowing astronomers to obtain radio pictures of galaxies that are sharper and more detailed than optical ones. The VLA is currently involved in research programmes investigating the properties of radio galaxies and quasars and the structure of stars.

A number of individual radio telescopes can be used together to act as one large telescope. The movable dishes each 25 metres in diameter which make up the Very Large Array in New Mexico can be combined to produce the equivalent of a 27-kilometre optical telescope.

It is also possible to use apparently unconnected radio telescopes as one instrument simply by sending the amplified radio signal from each component across the ordinary radio system to a control centre. In Great Britain seven individual telescopes spread across England and Wales from Jodrell Bank to Cambridge can be combined in this way to make the equivalent of a radio dish 133 kilometres across. They are jointly known as MERLIN (the Multi-Element Radio-Linked Interferometer Network). The size potential of this technique is limited only by the size of the Earth itself. The system known as the VLBI (Very Long Baseline Interferometry) set up in recent years links instruments on different continents to create a telescope equivalent to a radio dish with the diameter of the Earth.

4 · The Observatory

Astronomers work in many different surroundings, including the lecture theatre, the laboratory and the office, but it is the observatory which is undeniably the astronomer's own ground. In its very broadest sense an observatory is simply a place where astronomical observations can be made. At one end of the scale the definition includes the portable observatories used by the astronomers who accompanied Captain Cook on his three voyages of exploration in the 1760s and 1770s. To protect themselves and their instruments from the weather and the curiosity of the natives these men set up tent-observatories where Cook landed, which they removed once

their work was finished. In stark contrast is the Kuiper Airborne Observatory, named after the astronomer G. P. Kuiper, which started work in 1975. This sophisticated aeroplane can fly at an altitude of 12 kilometres and has a 91-centimetre reflecting telescope situated in its fuselage just ahead of the wings. The telescope has all the flexibility of a ground-based instrument, its orientation and focus being fully adjustable from the cabin. This flying observatory is used about eighty times a year for a few hours at a time.

Early astronomers tended to place their instruments on rooftops, as at Greenwich (right). Portable tent observatories, like those used by Captain Cook (left), were developed for expedition use.

The nature of an observatory and its location are dictated by the type of job it is expected to do. Captain Cook's tents were nothing more than a form of protection, whereas the whole point of an airborne observatory is that it can get above most of the Earth's atmosphere and so offer much better viewing conditions than could be achieved on the ground. Flying observatories are also mobile, giving the added advantage that they can sometimes be used to 'track' an object for a longer time than would be possible on Earth. An eclipse of the Sun, for example, is visible from a ground-based observatory for only about eight minutes, whereas an aircraft can follow the Moon's shadow as it moves across the Earth and so observe the effects of the eclipse for significantly

LONDINUM.

longer. From the Kuiper observatory, an eclipse could be kept in view for some 16 minutes, and if you were lucky enough to be flying in Concorde, which moves considerably faster, this time would be extended to a few hours.

Wherever an observatory is based it needs to have an unimpeded view of the heavens. The observatories built to house the first telescopes in the seventeenth century were often placed on a roof or in specially designed buildings erected on the top of a hill, such as the Royal Observatory complex constructed at Greenwich in 1675. At this date there were no street lights to affect what could be seen, and pollution from houses and factories had not significantly clouded the atmosphere. By the nineteenth century, however, viewing conditions in cities had deteriorated substantially and it was recognized that observatories should ideally be placed in remote locations, so the astronomer could escape the effects of man-made pollution. It was also realized that they should be at as great an altitude as possible in order to minimize the problems involved in viewing through the Earth's atmosphere.

The turbulent gaseous mantle which keeps us all alive is, in a sense, one of the greatest barriers to exploring the heavens. It not only affects the actual performance of a telescope, for example by gradually degrading mirror surfaces, but also distorts the light which reaches us from objects in the heavens. It is the atmosphere which makes the stars twinkle and blurs the images of galaxies. Placing observatories on top of mountains where the atmosphere is both clear and thin is one solution to the problem of atmospheric distortion. But it is even more satisfactory to site observatories in orbit beyond the Earth's obscuring atmosphere. As discussed in Chapter 6, developments in the past decade have brought this goal within reach.

Nevertheless the day of the traditional observatory with its distinguishing dome is by no means over. Just like Captain Cook's tent, these

The domes of the Old Royal Observatory at Greenwich. The dome of the altazimuth building in the foreground, which houses a 6.8-inch refractor, is opened by hand using a pulley system. The fully automated dome behind houses the 28-inch refractor.

intriguing white mushrooms protect the instruments, and are designed to be opened quickly and efficiently when a telescope is to be used, allowing the astronomer a view from horizon to horizon. The earliest observatory domes built in the nineteenth century were rested in a circular trough of large ball-bearings so they could be easily manipulated. A pulley system was used to open the dome to provide an aperture for the telescopes to look through and the whole dome was then moved around until the instruments were pointing towards the aperture and the open sky. This rather cumbersome arrangement meant that the observer had to leave his telescope every time the dome needed moving. Today domes are opened and moved by electricity and the astronomer can operate them without having to leave the eyepiece of the telescope or the control room nearby.

Inside the dome the telescope is supported on very solid foundations. Observatory telescopes, of massive proportions compared to anything you might use in your back garden, are extremely heavy and need to have a very firm base if they are not to be constantly moving out of alignment. In a very few examples, the observatory floor can be moved up or down with respect to the telescope so that the observer can reach its eyepiece comfortably. But not all telescopes are housed in domed buildings. The specialized instrument known as the transit telescope does not need this facility because it remains stationary, observing the heavens as they move in front of it through a slit in the observatory roof, usually revealed with a series of shutters. The transit telescope is never used to track objects as they move across the sky and so does not need access to a full 360° circle. The advantage of a telescope of this kind is that it can be used to measure the positions of the stars with great accuracy.

Until a few decades ago nearly all the world's major observatories were situated in the northern hemisphere and concentrated on the exploration of the northern celestial sky. Astronomers had ventured to southern latitudes and charted the heavens and some permanent observatories had been set up, such as the Royal Observatory established in 1829 at the Cape of Good Hope in South Africa. But, compared with the northern hemisphere, until the middle of the twentieth century the southern sky was still relatively unexplored. This may seem rather surprising in view of the fact that the southern

Hand-operated shutters open up a slit in the roof for the specialized transit telescope at Greenwich. This instrument defines 0° longitude, dividing east from west.

skies are particularly rewarding, giving views towards the centre of our Galaxy and of the Magellanic Clouds, the two companion galaxies to our own. The situation was remedied in the 1950s when a number of European countries co-operated to establish a major observatory in the southern hemisphere, leading to the foundation of the European Southern Observatory (ESO) in the 1960s. The site chosen was a 2400-metre ridge in Chile about 600 kilometres north of

Observatories on remote sites need a full range of support facilities. This view of the European Southern Observatory in Chile shows hotel, administration and workshop facilities in the foreground. There is also an airstrip close by.

Santiago and 70 kilometres from the Pacific coast. The shape of the ridge gave the observatory the name by which it is now most commonly known, La Silla (the saddle).

The site was chosen because observing conditions there are excellent. The air is very clear, dry and stable; the average rainfall is only 50 millimetres a year, there is little air turbulence and the sky is clear for over six consecutive hours on 255 nights a year. In fact, La Silla is a high altitude desert. Moreover, it is not affected by man-made pollution as there is very little settlement or industrial activity close by. When the proposed site for the buildings was approved, an area of land surrounding them was also acquired so as to ensure the observatory is not adversely affected

by future development, such as the light and dust pollution that would result from a new mining settlement. There is also very little difference between the day- and night-time temperatures here, another important factor in siting an observatory as problems are caused by instrumentation expanding and contracting in line with temperature fluctuations. Observatory domes are never heated and some are even cooled in order to keep the instrumentation within the dome at the same temperature as the night-time air. Observing can be a very uncomfortable and chilly business!

The countries initially involved in the ESO observatory were Belgium, West Germany, France, the Netherlands and Sweden, but they have since been joined by Denmark, Italy and Switzerland. Because the observatory is so remote a number of, perhaps unexpected, facilities have been built alongside the telescope domes, such as the airstrip without which it would take days rather than hours to reach the site. La Silla also includes a health centre, accommodation for the observatory's permanent staff and visiting astronomers and workshops where instrumentation can be repaired or altered quickly so no observing time is lost.

There is a wide range of instrumentation at La Silla, but most of the telescopes are optical. Specialized equipment of the type discussed in Chapter 5 is used with the telescopes to analyse the light they receive and give further information on the nature of the objects being observed. The most powerful instrument at La Silla is the 3.6-metre reflecting telescope which is used to observe faint and distant astronomical objects like quasars and galaxies. There are fourteen telescopes in all in operation, of which eight are financed by the ESO and the others sponsored by member countries.

One of the more exciting instruments is a 15-metre radio telescope called SEST (Swedish-ESO Submillimetre Telescope), which has been operational only since 1987 (see p. 29). This is the only

The southern sky offers astronomers the opportunity to look towards the centre of the Galaxy and our companion galaxies, the Magellanic Clouds. This view of the Small Magellanic Cloud was taken from the Anglo-Australian Observatory.

large radio telescope in the southern hemisphere capable of operating at submillimetre wavelengths (between one-third of a millimetre and a millimetre), i.e. at the extreme end of the infra-red range. La Silla is ideal for an instrument of this kind because the climate is so dry – it is usually water vapour in the atmosphere which absorbs submillimetre wavelengths. Moreover, from this southern hemisphere site it is possible to observe interstellar molecules in the inner regions of our Galaxy and the Magellanic Clouds which are not visible in the northern skies.

At about the same time that the ESO was being developed, the various governmental and independent organizations interested in astronomy in the United States were also pooling their resources to set up advanced observing facilities. Their scheme developed into what we know as the US National Optical Astronomy Observatory, which includes a number of individual centres under the overall umbrella organization. The first to be opened, in 1958, was the Kitt Peak National Observatory sited at 2100 metres in the Quinlan Mountains of the Sonoran Desert about 100 kilometres from Tucson, Arizona, where the atmosphere is again exceptionally clear and dry. This was followed by a southern hemisphere observatory in Chile, the Cerro Tololo Inter-American Observatory, and then by the dual-site National Solar Observatory based partly at the Kitt Peak site and partly at Sacramento Peak, New Mexico. There are fifteen major telescopes at the Kitt Peak site and almost that number of domes. The instrumentation includes the McMath solar telescope, the largest such instrument in the world. This observes our local star, the Sun, by day, in contrast to the more conventional telescopes which observe the heavens by night. The largest optical telescope on the site is the 4-metre Mayall, available for use by visiting astronomers for 60 per cent of the time it is operational.

Observing programmes at Kitt Peak are very varied. Recent work has included seeking evidence for the existence of black holes and other planetary systems, measuring the masses of galaxies and monitoring large-scale movements within them, such as the spin velocity of different regions of the galactic disc, and infra-red observations of the Milky Way. A 12-metre radio telescope, observing 24 hours a day, has been used to study the characteristics of star-forming gas clouds, helping astronomers to build up a picture of the evolution of the Universe. Unusually, this radio telescope is shielded by a dome to protect it from the high winds that buffet the site. The information gathered from this instrument is stored on magnetic tape which a visiting astronomer can take away for analysis at his home base, leaving the observatory facilities free for others to use.

This wintery view of the Kitt Peak National Observatory was taken from the catwalk of the dome housing the 4-metre Mayall telescope.

5 · Tricks of the Trade

One of the basic questions in astronomy is what planets, stars and galaxies are. What are they made of, where did they come from and how will they change in the future? These are the problems that professional astronomers are wrestling with, with new techniques continually being developed to broaden the horizons of their research.

Any new branch of astronomical research usually starts with a phase of 'exploration'. Just imagine what it would feel like to be the first person to train a telescope on the sky. Just think of all the new discoveries you would make and how they would come thick and fast at the beginning. A similar phase of exploration has happened every time a new technique has been developed, such as the instrumentation involved in monitoring X-rays, or infra-red and ultra-violet radiation. Detailed scientific investigation follows this exploration phase, using the information which has been collected. Each type of investigation has its own instrumentation. This may be used alone, but is usually supplemented by the traditional telescope and perhaps by other techniques as well. Some, such as photography, will be familiar to everyone; others, such as polarimetry, which measures the direction and movement of electric waves within a light beam, are more specialized, but all can reveal valuable information. Those looked at in this chapter can be used on Earth, or from satellites in orbit (see Chapter 6), or on space probes to investigate the Solar System (see Chapter 8).

Today, as in the past, *photography* has an essential role to play in astronomy, recording

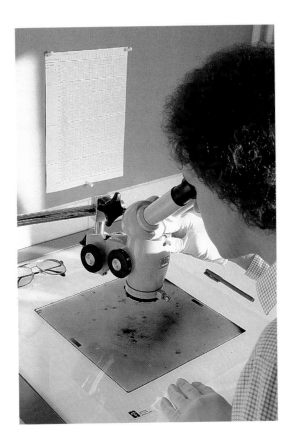

Photographs taken by astronomers using the world's best telescopes provide an accurate and detailed record of the night sky. Sky 'atlases' made from the original photographic plates are distributed to scientific institutes around the globe, and are an invaluable aid to astronomical research.

43

light levels to produce permanent images of, for example, a field of stars, a nebula, or the surface of a planet. Astronomical photography is one of the few remaining areas where glass rather than plastic film is used to support the photographic emulsion, as this eliminates distortion. The plates are often very large, those used with the 1.2-metre UK Schmidt telescope at the Anglo-Australian Observatory, for example, being 14 inches square and 1 millimetre thick. In contrast to normal photography, plates used in astrophotography are curved to match the curvature of the focal plane in which the image is formed, a result of the telescope's huge field of view, the corners of those used in the Schmidt telescope being bent forwards by 5 millimetres relative to the centre. The glass springs back flat again when it is removed from the plate holder.

Only very high speed film is used in astronomical photography as this enables very faint objects to be recorded with relatively short exposure times. Film speed is often considerably increased by baking the film at a temperature of 65°C in an atmosphere of pure nitrogen. Nevertheless, some astronomical photographs still need an exposure time of several hours, simply because the stars are so faint.

Photography has been invaluable in making surveys and maps of the sky, finely-tuned measuring devices allowing the positions of stars and other bodies, such as planets and comets, to be read off the plate with great accuracy. The Schmidt telescopes have been used extensively for this purpose, about 600 plates being needed to survey the southern sky.

The first deep-space photographs were taken in the 1880s. Colour photography of objects outside the Solar System only started in 1959, when William C. Miller experimented with fast colour film, using the 200-inch Mount Palomar telescope. Although this was the biggest instrument in the world at the time, it took an exposure of four hours to obtain a good colour image of the Crab Nebula. Today most colour photographs are obtained by colour separation, using blue, green and red filters to produce images of each colour 'band' and then combining these in the dark room to give a composite picture, in the same way as colour is reproduced in printing.

Various techniques can be used to enhance detail or isolate an area of sky. In unsharp masking the plate is printed through a low contrast contact print taken from the back of the plate so it is slightly blurred. Combining this blurred image with a sharp one gives better contrast in the final result, revealing amazing detail in the over-exposed parts of the original. In the technique known as photographic amplification, the film is processed so as to remove as much of the background as possible, thus giving a clearer image of the object being studied. The thin wispy shells surrounding many elliptical galaxies were discovered in this way.

Photography has also been used in space astronomy, the lunar orbiter spacecraft, for example, each carrying 80 metres of 70-millimetre film, which was developed on board. As the spacecraft were not designed to return from the Moon, the film was scanned in situ by an electronically controlled light beam which examined each photograph in strips 2.5 millimetres wide by 5.5 centimetres long. The information on each strip was then transmitted by radio back to Earth. In this way 1950 high resolution photographs were obtained, covering slightly more than 99.5 per cent of the Moon's surface. On the best of these, the detail was fine enough to identify objects only 1 metre across.

Both photography and what is known as *imaging* produce pictures in a form that seems familiar, but imaging is a largely electronic process, such as is used in television. Images of the surfaces of the planets and their satellites, such as the *Voyager* pictures of Jupiter and Saturn and *Mariner* images of Mars and Mercury, have provided some of the most startling results of the space age. Most have been produced by using what are basically television cameras. Two lens systems were used on the *Voyagers*, a wide-angle lens covering an area 90 kilometres square seen from a distance of 1600 kilometres, and another lens imaging a field of view only 11 kilometres across from the same distance. The exposure time could be varied between 0.005 and 15 seconds by manipulating a shutter. An electron beam takes 48 seconds to scan the image recorded by the camera, transforming it into a matrix made up of 800 lines each composed of 800 picture elements

Edgar Mitchell pans a television camera as he walks across the surface of the Moon in February 1971. Scientists on Earth used the transmitted images to advise the Apollo 14 *astronaut where he should walk.*

(known as pixels). Very subtle distinctions in tone can be picked up by these pixels, as the light intensity recorded by each one is registered on a scale from 1 to 256. This is then recorded by a computer, which transmits the information back to Earth incredibly fast. From the distance of Jupiter, for example, images were transmitted at the rate of 115,200 bits per second, so it took less than a minute to send back each picture.

The greatest recent advance in astronomical imaging was the development of the silicon chip known as a charge-coupled device (CCD) in the late 1960s. In contrast to the most sensitive photographic plate, which only registers about one in forty of the light particles falling on it, the CCD detects almost every one. It is also sensitive

This electronic chip, called a charge-coupled device, is much more sensitive to light than photographic emulsion. CCDs are now rapidly replacing traditional photographic plates in astronomical studies.

to a large range of wavelengths, stretching from the near infra-red part of the spectrum to X-rays, whereas the photographic plate picks up very little beyond the visual band.

Today all the world's large telescopes use CCD detectors and they have been used to image Halley's Comet from the *Giotto* spacecraft and will be used on board the Hubble Space Telescope (see p. 67). The CCD has the additional advantage that it can detect a large range in intensity, and so can accurately record both faint and bright objects at the same time. And the information it produces is in an ideal form for feeding into a computer for subsequent analysis. Unlike a photographic plate or a television camera, which respond in a complicated, somewhat logarithmic fashion to the amount of light that reaches them, the CCD gives a signal which is proportional to the true light level, in other words a more accessible form of information which is easier to use. Finally, CCDs are solid devices which are remarkably stable. The largest CCD to date has a sensitive array of 2048 × 2048 pixels on an area about 6 centimetres square (it also costs about £50,000).

Unlike photography and imaging, the technique of *spectroscopy* is not concerned with producing a picture of an object but with analysing its nature. As we have seen, all light can be broken up into its constituent colours. A rainbow is nothing more than the light from the Sun, our star, split up into the familiar colour sequence: red, orange, yellow, green, blue, indigo and violet. But the spectrum of visible light from a star can tell us much more than this. Variation in the strength of each colour band is directly related to the temperature of that part of the stellar atmosphere that is emitting the radiation. By analysing the energy curve of sunlight we can tell that the Sun has a surface temperature of about 5700°C. We can also tell what it is made of. Close inspection of the spectrum of sunlight shows that it is crossed by a series of dark lines (known as Fraunhofer lines after the German scientist who first discovered them in the early nineteenth century). The intensity of these lines and their position in the spectrum can be used to deduce the composition of the stellar atmosphere and its temperature and pressure.

The essential characteristic of a spectroscope is its ability to disperse the light it receives, breaking it up into the colours of the spectrum. In the case of the Sun, where there is an enormous amount of available energy, the spectrum can be many metres long and much fine detail can be resolved. But with very faint stars we have to be satisfied with spectra that are only a few millimetres in

A two-hour exposure with a CCD using the 4-metre telescope at the Cerro Tololo Observatory revealed a number of hitherto undetected galaxies, too faint to be recorded by photographic emulsion.

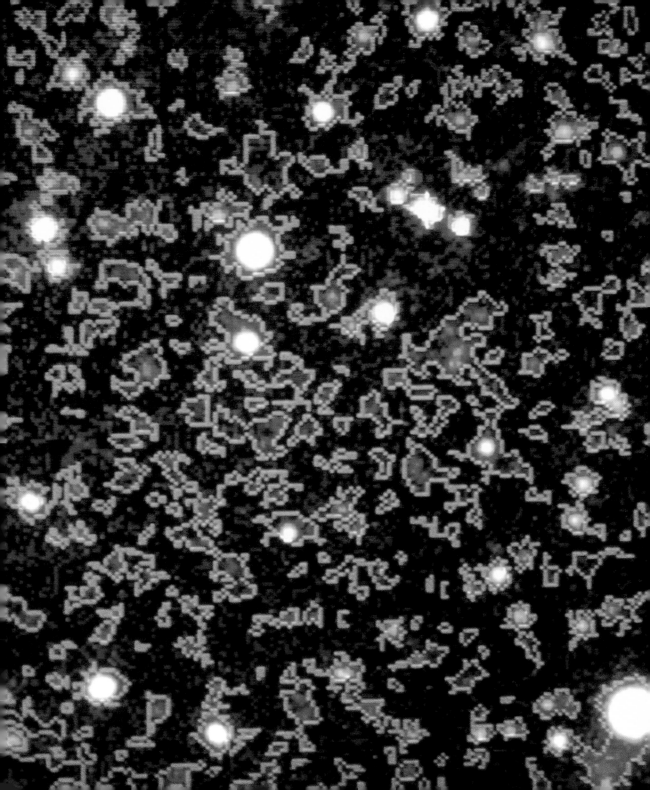

The spectrum of a star can be used to deduce information about its temperature and composition, the position of the dark vertical bands indicating the presence of particular elements.

length. Light is dispersed in a spectroscope either with a prism, usually of glass or quartz, or with a diffraction grating, a fine grid of lines which reflects the light in such a way that a spectrum is produced. An optical system is used to focus this spectrum as sharply as possible. The diffraction grating has the advantage that it loses very little light, whereas all prisms absorb light as well as transmitting it. The spectra can be recorded by allowing the light to fall on to a photographic plate, a vidicon (television camera), or a CCD.

Spectroscopy is of fundamental importance to astronomy as it can be used to measure many of the physical properties of a star, including its chemical composition. It was first used astronomically by Sir William Huggins in the late nineteenth century and he was responsible for making the rather astounding discovery that stars were made of exactly the same chemical elements as are found on Earth. Later work showed that slight differences in the composition of stars were related to the age of the star and were caused by nuclear reactions producing energy in its core, during which lighter elements, such as hydrogen and helium, were processed into heavier elements, such as carbon, oxygen and silicon (see *The Greenwich Guide to Stars, Galaxies and Nebulae*).

Unfortunately measuring the spectra of stars can be difficult because the amount of energy we receive from them is frequently very small. So a number of spectroscopes with different characteristics are needed to cope with the wide range of stars in the sky. The faintest stars whose spectra astronomers are trying to measure are about a hundred million times less bright than Sirius, the brightest star in the sky, and this again is about ten thousand million times fainter than the Sun.

RIGHT *A spectroscope fitted to the 4-metre Mayall telescope at the Kitt Peak Observatory is being adjusted at the start of an observing programme.*

Omega Centauri is the largest globular cluster in the Milky Way. Its unusual elliptical shape – most other globular clusters are round – shows clearly in this photograph. By studying the light from Omega Centauri through a spectroscope astronomers discovered the cluster is rotating and this could be the cause of its unique shape.

One interesting application of spectroscopy is in the measurement of velocity. This technique makes use of what is known as the doppler effect, most easily demonstrated by listening to the pitch of an ambulance siren as it approaches, passes and then travels away. The marked change in the sound is related to its velocity. Similarly, the wavelength of light from any particular source is directly related to how fast it is travelling and whether it is moving towards the Earth or receding away from us. By using the doppler effect as related to light, astronomers have been able to chart the expansion of the Universe, the rotation of our Galaxy, the orbits of binary stars, the pulsation of stellar surfaces and the spin rates of stars and planets. Spectra also reveal slight

49

changes in atomic energy levels caused by magnetic fields and thus enable the intensity of these fields to be measured. Sunspots, for example, have been found to have a very intense magnetic field, much greater than that which orientates compasses on Earth.

Photometry is another technique which astronomers use to investigate the essential characteristics of a star. A photometer is rather like a light meter, but one that can be adapted with filters to measure and record electromagnetic radiation from astronomical objects over a range of wavelengths, although it is usual to focus on a relatively narrow band. Depending on the selected wavelengths, usually designated by letters of the alphabet, a whole range of filters may be used. The standard practice is to record U, ultra-violet, B, blue and V, visual (which is actually green) radiation. Other filters will select wavelengths known as H, I, J, K and L which move off through the infra-red part of the spectrum.

Photometry is used to study regions of the sky and to investigate particular bodies of interest. Whole star fields are captured in photographic photometry; the relative brightness of the stars recorded through the different filters can then be used to assess their surface temperature and luminosity. Unfortunately a single photographic plate covers only a restricted range of brightnesses. In contrast, photoelectric photometry concentrates on a single stellar source, amplifying the energy received by electronic means so that very faint starlight can produce an easily measurable electrical signal.

Photoelectric photometers are very stable and the output can be easily fed directly into a computer. They have proved invaluable in following the light curves of variable stars and in investigating the composition of planetary surfaces, i.e. assessing what the rock and dust is actually made of. This is how methane ice was detected on the surface of Pluto. The photometer can also be combined with a polarization analyser to form a *polarimeter*, which can be used to measure the electromagnetic orientation of the light beam. Polarimetry is very useful in studying the planets and has given astronomers considerable information about the physical texture of the planetary surfaces, how rough the surface is and the size of the dust particles that cloak large areas. Some information on the composition of the surface can also be deduced. Even though we now have samples of Moon rock on Earth, these were collected from a very limited area and analysing them alone would be similar to, for example, drawing conclusions about the Earth from samples taken just from the Sahara Desert. Earth-based polarimeters have given scientists information about a much larger area of the Moon's surface to add to that obtained from the lunar missions.

Another way of obtaining pictures and spectra from very, very faint stars is to use an *image intensifier*, an electronic device which brightens up complete images obtained from observations at infra-red as well as visual wavelengths. The enhanced image appears on a thin sheet, known as a phosphor, where it is about fifty times brighter than the original image. This enhanced image has the advantage that it can be photographed. If an electron multiplier is also used the image can be made up to a million times brighter. In this way astronomers are almost able to see in the dark.

Astronomers also obtain fine detail through *interferometry*, the technique described in Chapter 3 which involves combining the light, or radio waves, from two or more telescopes observing the same object. Alternatively, information obtained from an array of telescopes in this way can be

This photograph of the region adjacent to the red supergiant Antares in the constellation of Scorpius was obtained with the 1.2-metre UK Schmidt telescope at the Anglo-Australian Observatory. It was made by combining three black and white images taken through different colour filters.

used selectively to focus on a particular feature or characteristic. Interferometry has proved particularly valuable in radio astronomy; slight time delays in the receipt of radio signals at each of the component telescopes can be used to pinpoint the precise position of a radio source.

As mentioned in Chapter 3, the largest purpose-built array of telescopes which can be combined to simulate one large aperture is in New Mexico (see p. 34). The 27 radio telescopes here, known as the VLA, have been used to produce very detailed maps of the radio sky, maps which cover an area ten times larger than those produced with a typical large optical telescope. Another intriguing instrument is the optical interferometer at Narrabri in New South Wales, Australia, where two reflecting telescopes can be moved freely round a circular track with a radius of 90 metres. The combined image produced from these two reflectors can be used to assess the shape of a star and how light output varies across its disc. The Narrabri interferometer can also be used to measure the size of near giant stars, and has revealed 'starspots' like the more familiar sunspots.

Another type of interferometry relies on combining a large number of short-exposure observations of a star or planet taken one after another. These are then fed into a computer which is programmed to use the information selectively to produce the best possible image.

Analysing the information from this sophisticated equipment would be much slower and much more difficult if astronomers were not backed up by *computers*, the everyday workhorses of any observatory. Computers are used to control large telescopes, directing the motors that swing an instrument round to point at the part of the sky the astronomer wants to observe. With their aid, moving objects like asteroids and comets can be carefully tracked. Of equal importance is the vital role computers play in analysing all types of astronomical information, such as the data in a CCD image, and in helping

This radio image of Saturn obtained from the Very Large Array telescope shows the cooler rings girdling the planet. These appear blue on either side but yellow where radiation coming from Saturn affects our view.

astronomers to make sophisticated, intricate calculations, such as the exact paths of the planets around the Sun, taking into account all the minor perturbations introduced by their gravitational attraction for one another. A computer also enables astronomers to move both forwards and backwards in time, and to do so speedily, producing information on, for example, where the planets were on 7 November 4000 BC in a matter of minutes, a calculation that an individual would probably need weeks to complete.

The computer's ability to tackle even more complicated mathematical problems, such as how pressure, temperature and density change moving towards the centre of a star, has revolutionized research, enabling the astronomer to do in weeks what would previously have taken him a lifetime. With computers, large telescopes, detectors that can pick up every photon of light, instrumentation that can travel to other planets as well as view the sky from space and collect information in all wavelengths, the present-day astronomer is remarkably well equipped.

Even so, two more traditional astronomical instruments still have a vital part to play, and are still regularly used. A device known as the *micrometer*, invented soon after the telescope, enables the distances between stars to be measured when it is fixed to the eyepiece. It has been used, for example, to assess the movement of visual binary stars orbiting their common centre of mass, enabling astronomers to measure how far apart the stars are, the angle at which we see them, and how these change in relation to the phase of the orbit (which may take many years to complete). This information can be used to assess the mass of each star. Micrometers are also used to measure the positions of stars, planets and

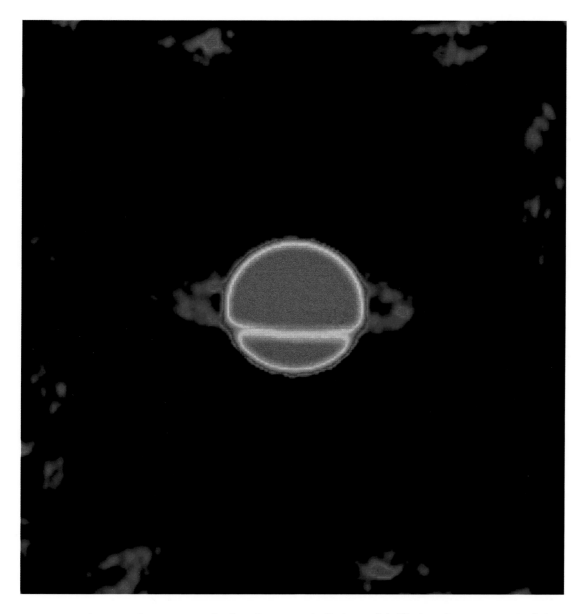

comets on photographic images. Such information helps astronomers to calculate comet orbits and the distances of the stars. The *clock* is similarly useful. Timing the movement of stars across the sky, for example, enables their celestial co-ordinates to be measured.

6 · *Astronomy above the Clouds*

Analysing light and radio waves has given us a good picture of our space environment, and astronomers continue to use conventional and radio telescopes on Earth to investigate the heavens. But, as the diagram on p. 11 shows, we receive many other kinds of electromagnetic radiation from the heavens and it is only by investigating them all that we can hope to build up a complete picture of the Universe and its evolution. The problem is the Earth's atmosphere, which effectively absorbs most kinds of electromagnetic radiation. It is only by devising instruments to go above the clouds that we can hope to get an X-ray or ultra-violet picture of the Universe, images that suggest new ways of looking at the optical universe, add new dimensions to familiar objects and reveal hitherto invisible phenomena. The results are as spectacular as those we would get if human beings were suddenly given an X-ray eye, which could see through people's clothes and bodies to reveal their skeleton in all its glory.

One of the first types of astronomy above the clouds to be developed and one that still yields spectacular results is *X-ray astronomy*. Celestial X-rays were first observed on 5 August 1948, when the darkening of photographic emulsion carried in a V2 rocket was taken to infer the existence of solar X-rays. This finding was confirmed when another V2 rocket detected intense solar X-radiation, but surprisingly a detailed X-ray picture of the Sun showed that X-rays accounted for only one millionth of the total energy emitted by our parent star. About ten years later other possible sources of X-rays, such

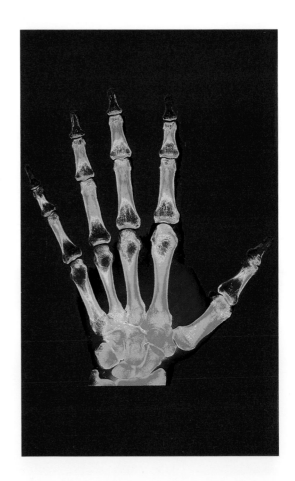

Just as an X-ray image of the human body is startlingly different from an ordinary photograph, so an X-ray picture of the night sky reveals a new and exciting cosmos.

54

The large cluster of galaxies known as Abell 1060 in the constellation of Hydra is a source of X-radiation. The X-rays come from a thin, hot gas at about 10 million °C in the inner part of the cluster, not from the galaxies themselves.

as supernova remnants and very hot stars, were identified. Specialized instrumentation developed and launched aboard rockets in the early 1960s detected the expected emissions and also identified background radiation from the cosmos.

The first X-ray astronomy satellite was launched on 12 December 1970 from a modified oil platform off the coast of Kenya. It was known as Uhuru, the Swahili for freedom, as it was launched on Kenya's independence day. Uhuru made a number of observations which would have been virtually impossible from rockets, its detectors scanning the same part of the sky many times over as the satellite slowly revolved. By the time the mission was completed four years later, Uhuru had identified some 270 X-ray sources, increasing the number known from 30 to 300. Because the satellite had been able to make a series of observations of each one, it also yielded information on variations in the intensity of X-ray emission and enabled the sources to be pinpointed more accurately.

The first UK X-ray astronomy satellite – Ariel V – was launched in October 1974. It spent five years in orbit and identified many more new X-ray sources, including SS433, a star system which sprays out gases at a quarter the speed of light and is one of the most unusual astronomical subjects ever studied. The Einstein Observatory satellite in use from 1978 to 1981 went on to look at background radiation emitted in the region of distant galaxy clusters and at X-ray emitting gas in the remains of exploded stars. It also showed that strong X-ray emissions from the areas around stars is the norm rather than the exception, proving our Sun to be an abnormally faint X-ray source, and discovered many hundreds of X-ray binary stars and hundreds of extragalactic X-ray sources.

The specialized instrument aboard this satellite was the first X-ray telescope with a sensitivity that matched the best Earth-based instruments used for optical and radio observations. When X-rays fall on to the mirror of a conventional telescope they simply pass straight through, but if they strike a surface at a very shallow angle they will be reflected off. The instrument carried by the Einstein Observatory had been devised to take advantage of this property, with concentric arrays of curved tubes to intercept and focus X-ray emissions. The inventor claimed that the advances made with his telescope over a thirty-year period were equivalent to those achieved in optical astronomy over 300 years. But even despite the enormous amount of information received from the Einstein Observatory, its X-ray telescope only succeeded in covering some 5 per cent of the entire sky, leaving the possibility that there were many X-ray sources yet to be discovered.

Many more observations were made by the European Space Agency's Exosat satellite, launched on 26 May 1983. Because it took almost four days to complete one orbit of the Earth, Exosat got long uninterrupted views of X-ray sources. Moreover, the Exosat programme was designed to be flexible. The path of the orbit enabled ESA's ground station at Villafranca near Madrid in Spain to keep the satellite constantly in view. The data received here was instantly relayed to the

Exosat – the European Space Agency's X-ray satellite – undergoes pre-launch tests to ensure it will perform perfectly when it is above the clouds.

control centre at Darmstadt in West Germany, where scientists could quickly assess how their projects were going and make changes if necessary, planning a new observation and programming the satellite accordingly within a few hours. During the three years it was in operation Exosat was reprogrammed to make about fifty unscheduled observations after scientists around the world had alerted Darmstadt to new potential X-ray sources.

Exosat stopped operating during the early hours of 10 April 1986, having spent the majority of its working life making observations. Only about three weeks in the entire three years had been lost because of problems connected with the spacecraft itself, or because of interference with the signals received due to increased solar activity. In all 1780 separate observations were carried out. Nine-tenths of the observing time was taken by the ESA countries; scientists from Great Britain took the lion's share, but those from Germany and the Netherlands were also granted large allocations. Although the satellite is no longer sending back information, there is still much work to be done on that already received, which is freely available to astronomers from all over the world. As well as discovering new X-ray sources, the Exosat instruments also recorded sources where the energy output varied rapidly and identified and assessed the strength of X-ray emissions. There are now plans to cover the sky in a detailed survey with the Rosat X-ray satellite, a German project jointly funded with the United States and Great Britain.

X-rays are, in general, emitted from the hot regions of space usually associated with upheaval and change and so present a more turbulent picture of the Universe than images in visible light. In optical wavelengths the Sun is a uniformly shining sphere with radiation emerging equally in all directions. The dark areas known as sunspots appear as black blemishes on the disc. But in X-ray (and radio) wavelengths the picture is reversed, the 'active' sunspot areas showing up

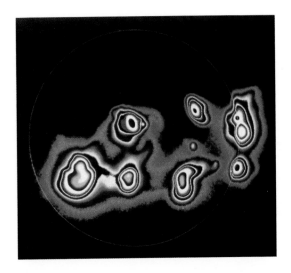

This X-ray view of our local star, the Sun, clearly shows the regions of most violent activity. These would not show up in optical light.

most brightly. X-ray images of the Sun also show up the corona, or outer atmosphere, only visible to the unaided eye during a total eclipse.

Objects too small to emit much light shine brightly at X-ray and radio wavelengths. The pulsar at the centre of the Crab Nebula gives off weak X-rays as well as radio waves, indicating that it is incredibly hot, with a temperature of some 2 million°C. Similarly, X-ray astronomy has revealed further details of the supernova remnant Cassiopeia A, showing that its outer layer is composed of interstellar gas swept up after the explosion at a temperature of about 50 million°C, and that an inner, cooler shell contains most of the matter from the exploded star itself, at a mere 10 million°C.

X-ray astronomy has also added to our picture of the curious objects known as quasars. Although quasars were first discovered with radio telescopes, they are also sources of intense X-radiation, 3C 273 proving a million times more powerful as an X-ray source than the Milky Way.

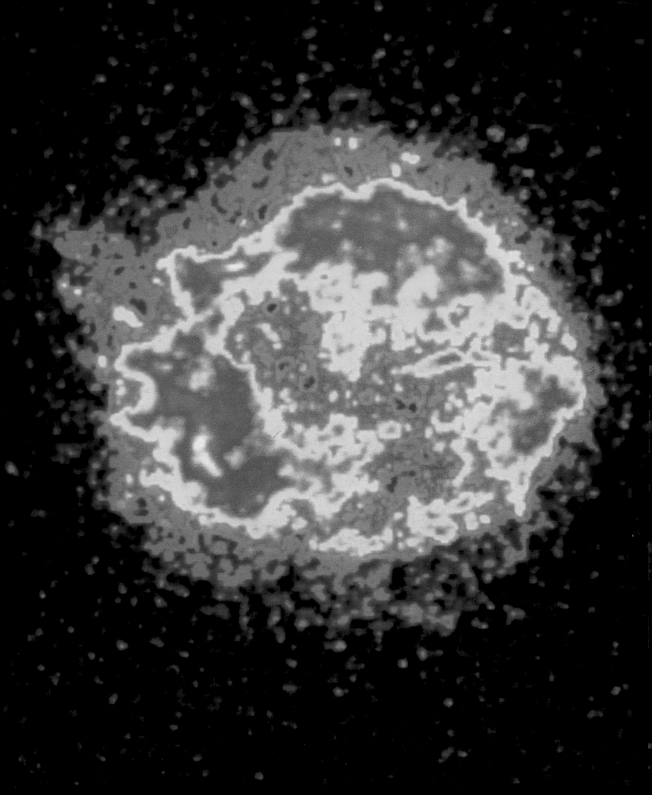

Moreover, whereas not all quasars emit detectable radio waves, X-rays have been recorded from all those – numbering hundreds – observed by the Einstein Observatory.

Even shorter than X-rays are *gamma rays*, which come at one end of the electromagnetic spectrum. These deadly rays have the ability to penetrate metals, for example a few centimetres of lead, as well as the soft tissues of the human body, so it is fortunate they are largely absorbed by the Earth's atmosphere.

Some gamma-ray observations can be made from high-altitude balloons, as the shortest gamma rays can penetrate the upper atmosphere to an altitude of about 10 kilometres, but most observations are made from satellites. Although some progress was made in the 1960s, the first real gamma-ray telescopes were not launched until the 1970s when the necessary techniques had been developed. Unlike other forms of electromagnetic radiation, gamma rays cannot be focused by reflection because they are smaller than the atoms which make up a mirror. Astronomers have to build up a gamma-ray image by using the direction of the rays to plot the position of the source in the sky. The specialized telescope launched in 1972 aboard the American SAS-2 satellite took readings for seven months in this way, and the instrument aboard the European COS-B satellite, which returned information from 1975 to 1982, was also of this type. Between them, these two projects have identified most of the thirty or so gamma-ray sources now known.

The supernova remnant Cassiopeia A, which appears as dimly-glowing gas on optical photographs, shines brightly at X-ray wavelengths. This view clearly shows concentric shells of gas centred at the point of the supernova explosion at the heart of Cassiopeia A. The strongest X-ray emission is denoted by the red and yellow ring; the blue and green areas are less energetic.

The gamma-ray sky which COS-B mapped is dominated night and day by the Milky Way. It is fascinating to see that the Sun is totally invisible at gamma-ray wavelengths unless there is a sudden outburst of energy in a solar flare. Generally, the brightest gamma-ray regions are those where cosmic gas is most dense, such as the Orion Nebula. The brightest star in the gamma-ray sky, discovered by Australian radio astronomers in 1968, is the Vela pulsar, which lies at the heart of the Vela supernova remnant. This is thought to have become a pulsar some 11,000 years ago, the second oldest known example. It spins 13 times per second, although observations have shown that it is gradually slowing down. Astronomers have only just begun to explore the gamma-ray Universe and it is hoped that more sensitive detectors currently being planned, such as NASA's Gamma-ray Observatory (GRO) to be launched in the 1990s, will reveal exciting new information.

Human beings cannot sense X-rays or gamma rays, but we feel *infra-red radiation* as heat, as anyone who has stood next to a fire will know. Everything at a temperature greater than absolute zero is a source of infra-red radiation – ourselves, a table, a telescope – although to different extents. This Earth-based radiation has often proved a problem when astronomers in observatories tried to record the weaker signals from space. Those celestial signals that do reach Earth are best detected with instruments in mountain-top observatories, but even there they are competing with the infra-red radiation from the atmosphere. In order to eliminate this unwanted interference, the strength of infra-red radiation from a particular object has to be constantly compared with the intensity of signals from an adjacent area of empty sky. Moreover, as most of the infra-red radiation from cosmic sources is absorbed by carbon dioxide and water vapour in the Earth's atmosphere, it is not surprising that infra-red astronomers want to get their instruments out into space.

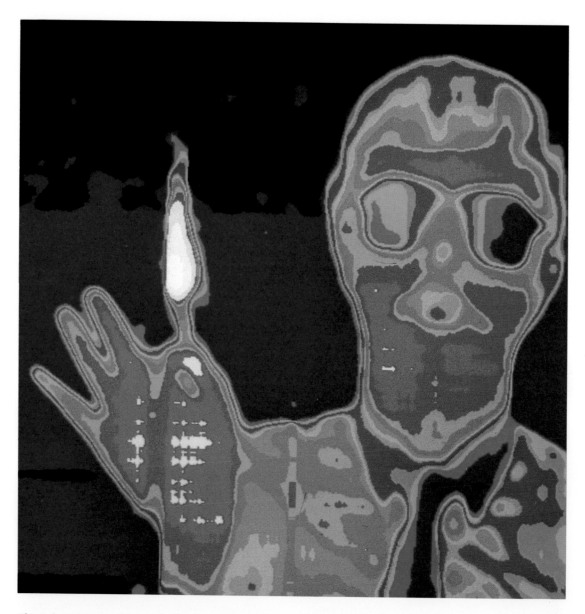

This infra-red image of a man holding a burning match shows how infra-red wavelengths can be used to detect sources of heat in the Universe. The white and red areas are the hottest, the man's glasses showing as two cool blue and green patches.

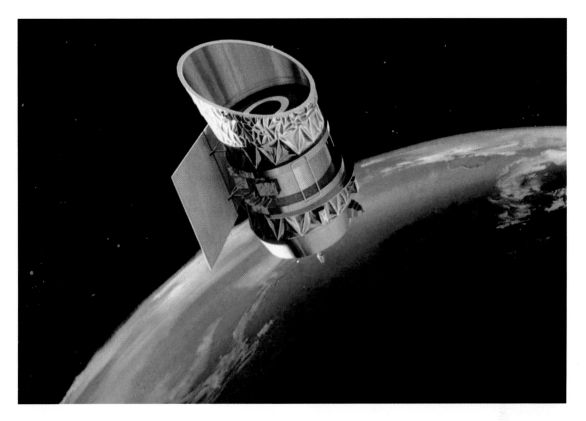

On each of the 300 days the infra-red satellite IRAS was in operation it discovered more infra-red sources than had been known before it was launched.

Infra-red radiation was first discovered by Sir William Herschel in 1800 when he was looking at the spectrum of the Sun produced through a prism. When he held a thermometer beyond the red end of the spectrum he detected an increase in temperature, the first record of infra-red radiation from our nearest star. In the next hundred years infra-red radiation was detected from the Moon and at the beginning of this century infra-red observations were made of some of the planets and a few bright stars. But it was not until the 1960s that infra-red astronomy really moved forwards, when the Americans Gerry Neuge-bauer and Bob Leighton used a ground-based telescope to complete the first major infra-red survey of the sky and discovered almost 6000 sources of infra-red radiation. High-altitude balloons and high-flying aircraft carrying infra-red telescopes (specially cooled to reduce the problem of heat from the instrument itself) have also been used, in particular on the Kuiper Airborne Observatory. But unwanted infra-red radiation interferes with the detector even at aircraft altitudes and the only satisfactory solution is to get a cooled telescope out into space itself.

The Infra-Red Astronomical Satellite (IRAS), a

joint American, Dutch and UK project, was launched in January 1983. The sensitive 60-centimetre reflecting telescope on board was cooled to −273°C, a temperature at which it was just about capable of detecting the heat from a crowded American baseball stadium from the distance of London. But this sensitivity only lasted while the cooling mechanism – a vessel containing about 500 litres of liquid helium – was fully operational. After 300 days all the coolant had evaporated and the telescope and detectors warmed up, consequently losing their sensitivity to infra-red radiation. Nonetheless, on every one of the 300 days that IRAS was working as planned, the satellite discovered more infra-red sources than the total known before its launch. About 180 days in total were used on an all-sky survey; during the remainder IRAS focused on objects of particular interest.

IRAS's ground station at the Rutherford Appleton Laboratory in England was deluged with information, receiving the equivalent of the complete *Encyclopaedia Britannica* twice a day, amounting to some 700 million bits of image data. Objects that had moved between observations showed up clearly, with five comets and a number of asteroids being added to those already known. The first comet to be found in this way – IRAS-Araki-Alcock – was named after the Japanese and British astronomers who simultaneously observed it on 26 April 1983. IRAS also showed up additional dust in the Solar System and faint wispy clouds beyond the orbit of Pluto – termed infra-red cirrus – as well as tens of thousands of other galaxies. But intriguingly it did not reveal any evidence for an additional planet in the Solar System, the so called 'planet X' which some astronomers believe is necessary to explain peculiarities in the orbits of the outer planets (see *The Greenwich Guide to the Planets*).

What IRAS *did* do, however, was to produce some evidence for planetary systems forming around other stars. The prominent young star Vega, for example, appeared unexpectedly bright in the infra-red, and was found to have a ring of gravel-sized dust particles surrounding it out to a distance comparable to the orbit of Saturn. IRAS recorded similar results in connection with many other young stars, and it might be that these rings of dust represent an early stage in the formation of planets, just as the Solar System planets are thought to have condensed from dust and gas orbiting our own Sun. As scientists are still working through the enormous amount of data returned, detailed analysis of the information from IRAS will take years to complete. No doubt there are still surprises to come.

Infra-red radiation is particularly useful for detecting the dust clouds associated with very young and very old stars, as these shine most brightly at infra-red wavelengths. When the contraction of gas and dust to form a new star

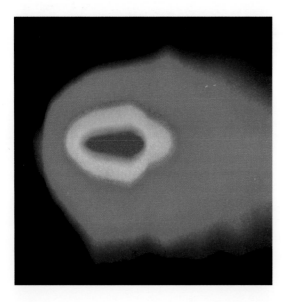

An infra-red image of the comet IRAS-Araki-Alcock obtained from the IRAS satellite shows the warm dust cloud around the icy central nucleus, whose ices are being evaporated by the heat of the Sun.

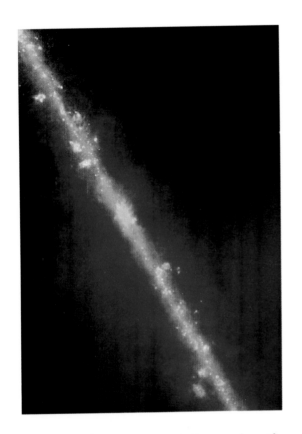

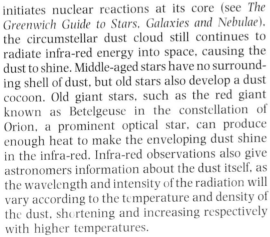

initiates nuclear reactions at its core (see *The Greenwich Guide to Stars, Galaxies and Nebulae*), the circumstellar dust cloud still continues to radiate infra-red energy into space, causing the dust to shine. Middle-aged stars have no surrounding shell of dust, but old stars also develop a dust cocoon. Old giant stars, such as the red giant known as Betelgeuse in the constellation of Orion, a prominent optical star, can produce enough heat to make the enveloping dust shine in the infra-red. Infra-red observations also give astronomers information about the dust itself, as the wavelength and intensity of the radiation will vary according to the temperature and density of the dust, shortening and increasing respectively with higher temperatures.

Using infra-red wavelengths astronomers have been able to record the centre of our Galaxy (left), a view that is obscured by clouds of dust (right) when observed in visible light.

Infra-red telescopes can also 'see' through clouds of gas and dust which obscure the view from optical instruments. One of the most exciting such regions is the Orion Nebula, where the denser central core contains enough gas to make 10,000 stars like our Sun. In the same way, infra-red telescopes have been used to penetrate the gassy clouds surrounding Venus and Jupiter. Temperature maps of Venus' lower atmosphere have revealed the presence of large-scale circulation movements which warm areas which are not receiving the direct heat of the Sun. Jupiter's infra-red radiation, in contrast, is concentrated in two belts parallel to the planet's equator which correspond to the brown belts seen in optical images.

Cooled detectors in space are also needed to explore *microwave radiation*, the short-wavelength radiation between the infra-red and radio part of the spectrum. This was first detected in space in the 1960s when the Americans Arno Penzias and Robert Wilson found that their sensitive radio aerial was picking up background radiation that seemed to be coming from all parts of the sky. This diffuse background radiation, thought to be a relic from the early Universe, is planned to be measured and mapped by NASA's microwave satellite, the Cosmic Background Explorer (COBE). It is hoped that COBE will provide data that will give us information about the evolution of the Universe, whether it is rotating and if it is expanding uniformly, how turbulent it is and clues as to how galaxies were formed and how and why they have come to be clustered. With the answers to these questions we could go some way to predicting the future of our Universe.

One form of short-wavelength radiation – *ultra-violet radiation* – was discovered as long ago as 1801 by the German Johann Ritter. It is extremely familiar today as the radiation that produces suntans, although fortunately most of it is absorbed by the upper ozone layer as otherwise we would all be badly burnt whenever the Sun shone. Like Herschel, Ritter made his discovery by using a prism to produce a spectrum of the Sun. When he held a piece of paper soaked in silver chloride beyond the violet end of the spectrum, the paper darkened, indicating invisible radiation at that wavelength. As some ultra-violet radiation does penetrate our atmosphere it was possible to get ultra-violet pictures of space from the 1940s, before the birth of the satellite era, by using rockets and high-altitude balloons. But once ultra-violet satellites were

launched in the 1960s, these crude pictures were rapidly replaced by more sophisticated images. A series of satellites launched by NASA in the 1960s and 1970s known as the Orbiting Solar Observatory gave detailed ultra-violet and X-ray pictures of the Sun. This information was augmented by the European TD-1 ultra-violet all-sky survey satellite, launched in 1972, which catalogued over 30,000 objects.

By far the most spectacular results have been produced from the International Ultra-Violet Explorer (IUE) satellite, launched in January 1978. Initially a British project under astronomer Bob Wilson, the satellite was subsequently jointly funded and run by NASA, the UK and ESA, and its facilities were therefore open to astronomers from the United States, the UK and Europe. Like Exosat, IUE's orbit was chosen so that it is always in contact with either NASA's control centre at the Goddard Space Flight Center or with the European control centre at Villafranca in Spain. This gives scientists working with the satellite the flexibility to guide and update their observing programmes while IUE is in action, a degree of control only previously available to optical and radio astronomers. IUE was only expected to be in operation for three years, but it is still going strong ten years after it was launched and still sending back information about the ultra-violet Universe. IUE has observed the ultra-violet spectra of nearly all the planets and their moons as well as looking at more distant objects – stars, nebulae, galaxies and quasars. IUE's observations of galaxies have helped scientists deduce what kind of stars they contain and the rate at which new stars are being formed. IUE has also detected a hot gaseous halo surrounding our own Galaxy, the Milky Way, and has been used in conjunction with ground-based optical and infra-red observations to give us highly detailed information on novae explosions and the supernova of February 1987.

Ultra-violet radiation is of particular interest to astronomers studying galaxies, as the strongest

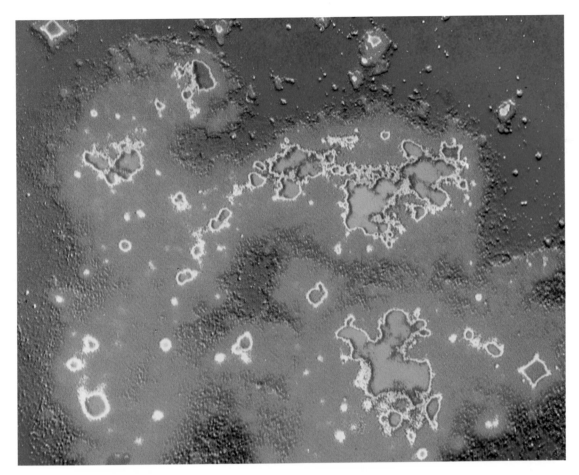

sources are hot, young stars such as those forming within the arms of spiral galaxies. An optical image of a spiral galaxy will be dominated by the brightly-shining older stars at the centre, but the ultra-violet picture emphasizes the spiral arms full of younger stars. In our own Galaxy, ultra-violet telescopes can detect hot, young stars in the Orion Nebula where these are not obscured by dust.

Ultra-violet techniques are also used for analysing the composition of stellar and other sources, as some common elements such as

An ultra-violet view of the Large Magellanic Cloud, the nearest galaxy to our own, captured by the Apollo 16 astronauts when they journeyed to the Moon in 1972. The image reveals young stars in the outer regions of the Galaxy, whereas an optical view would emphasize the old stars at its heart.

carbon and nitrogen show up more strongly in the ultra-violet than the optical spectrum. The Crab Nebula, for example, was shown to contain very little carbon in this way. The most easily studied source of ultra-violet radiation is the Sun.

65

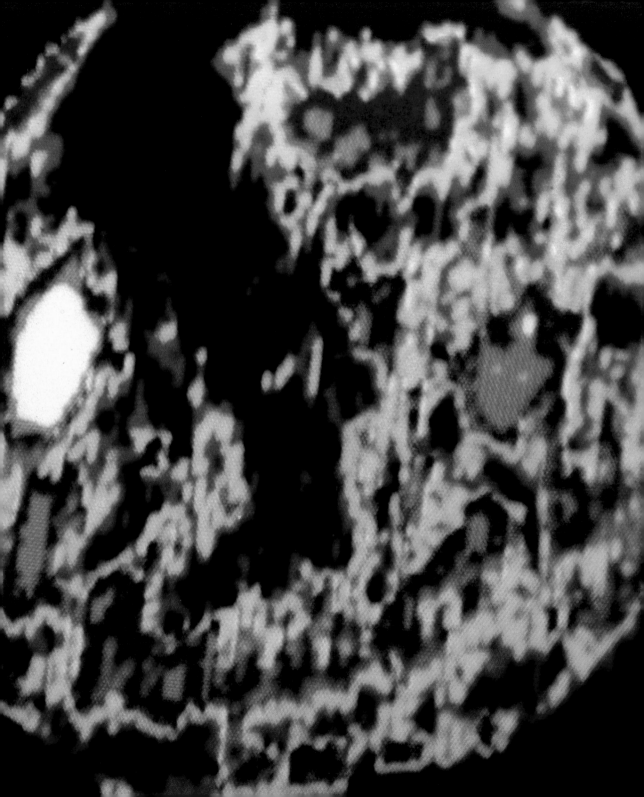

Photographing the Sun in ultra-violet light reveals holes in the corona, like that resembling southern Italy shown here.

When it was observed by Skylab for almost six months in 1973, ultra-violet techniques revealed that the corona is not a uniform atmosphere as had been previously thought, but contains large holes through which gas moves into space.

Observations with conventional astronomical instruments also benefit enormously from being made above the obscuring mantle of the Earth's atmosphere. The recently-developed Hubble Space Telescope (HST), a conventional Gregorian reflector telescope with unconventional capabilities designed to be launched by the American Space Shuttle, will offer the sharpest-yet images of celestial objects when it is in orbit at an altitude

Photographic plates used to survey the sky are invaluable aids in astronomical research. The arms of this spiral galaxy – NGC 6872 – have become 'unwound' due to the gravitational influence of a nearby galaxy (the small black oval above). Examining other plates will reveal if there are other examples of this phenomenon.

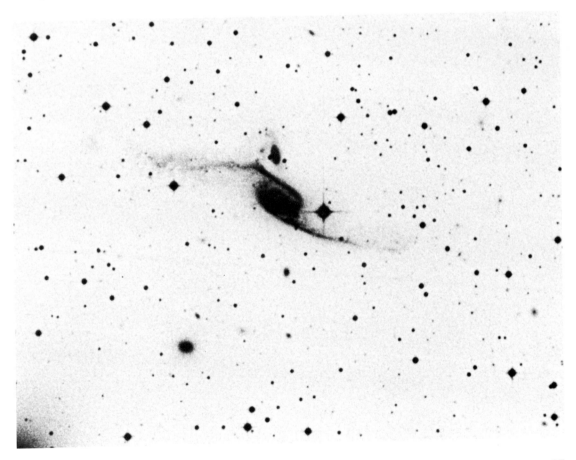

of about 500 kilometres. The optical surfaces of the 2.4-metre primary concave mirror and the smaller 0.3-metre convex secondary mirror mounted 4.6 metres in front of the primary are as perfect as modern technology can make them. The resolution attainable with this telescope will be ten times better than that achieved by comparable ground-based telescopes, which should ensure much more detailed observations of familiar and as yet unexplored objects. The HST will also enable astronomers to see seven times further into space than is now possible and to see objects up to fifty times fainter than anything seen to date. This revolutionary telescope will observe a volume of space 350 times greater than we are able to scan from Earth and it will function for about 10 hours a day, much longer than is possible on our planet.

The advantages offered by the HST compared to Earth-based optical telescopes can be likened to the difference between Galileo's first telescope and the human eye. But we still have some time to wait before we can see the pictures returned from this satellite. The HST is ready, but its launch has been delayed because of technical problems with the Shuttle programme as a result of the catastrophe in 1986 when seven astronauts lost their lives. Only when these are resolved will we know if there are surprises in store for us, and if the telescope will make revelations beyond our expectations and imaginations just as Galileo's telescope did over 350 years ago. Apart from the telescope itself, the HST is equipped with a mixture of other scientific instruments aimed at collecting a wide variety of information, such as making very precise measurements of the positions of stars and galaxies, gathering spectrographic data, and detecting extremely faint objects.

The HST will observe our Solar System as well as deep space. It will study each of our planets in turn and also be used to search for evidence of the existence of other solar systems, concentrating particularly on the 10 Sun-like stars less than 10 light years away and another 37 stars within 15 light years of our own Solar System. There are also a further 100 to 500 nearby stars which could have planetary systems. The telescope will not be able to make out individual planets if these exist, but it will be able to record perturbations in the motions of their parent stars that could indicate the presence of orbiting bodies like those in our Solar System. The development of stars will also be studied by looking at examples in all stages of their life-cycle, from the very youngest to the gas cloud remnants of supernovae. The faint-object camera will be particularly suited to working on globular clusters, in which it should be able to detect and record many white dwarfs, the compact, highly dense bodies which represent the final stage of most stars. By studying their size and luminosity it is hoped to piece together a more accurate picture of star death. It is also hoped that the HST will resolve some of the current questions about quasars. Pictures of these puzzling objects from ground-based instruments are too fuzzy to determine whether they are surrounded by the diffuse light of a galaxy, and it is hoped that the HST will give us the first high-resolution images and also the information to assess whether the quasar stage is associated with young or old galaxies.

Satellites in orbit over the Atlantic and Pacific Oceans will relay the data from HST to the Goddard Space Flight Center, from where it will be passed for processing and analysis to the Space Telescope Science Institute at Johns Hopkins University, Maryland. As the European Space Agency is covering 15 per cent of the costs and European astronomers will have 15 per cent of the observing time, there will also be a European data-analysis centre in Munich. An on-board computer will control the operation of the HST and handle the flow of data. The telescope is expected to last fifteen years, although it could be operational for considerably longer, especially if the Space Shuttle is used to extend its lifetime by providing regular servicing (see p. 76).

7 · *Astronomers in Orbit*

Even better than putting scientific instruments into orbit is to send up astronomers and other specialists in relevant areas, such as physics and geology, to look after them. NASA was already moving away from the pilot-trained astronaut of the 1960s and early 1970s when they sent the geologist Harrison 'Jack' Schmitt to the Moon on board *Apollo 17* in 1972, giving him the distinction of being the first scientist in space. Yet that trip was a one-off, only concerned with going to the Moon, collecting data and samples and returning. How much more useful it would be to have a base in orbit in space where scientists can have a laboratory and use their own instruments to carry out their observations and experiments. Such a base could also be used as a stop-over on the way to explore another planet.

The first steps towards this goal for American and European scientists were made with the introduction of the Space Shuttle programme. The Shuttle has already transported scientists into space to work aboard the space laboratory Spacelab, which is in orbit for limited periods of time, and NASA is now actively pursuing its aim of using the Shuttle to build an orbiting space station. The space station will be a permanent manned craft about the size of a football field, where scientists can live for long periods of time.

The Soviets are following a similar programme. The first really successful Soviet space station in orbit was *Salyut 6*, which had three telescopes on board. The largest was designed for recording radiation in the submillimetre, infrared and ultra-violet wavelengths. *Salyut 7*, launched in 1982 with an X-ray telescope on

The Space Shuttle transport system features strongly in American plans to establish a permanent, manned, orbiting station, one of its advantages being that the craft are reusable. Here the first Space Shuttle, Columbia, *lifts off on 12 April 1981, crewed by John Young and Robert Crippen.*

69

board, was similarly designed as a base for carrying out some astrophysical experiments. In an extension of the work of the *Salyut 6* crew, French-built cameras were used to study the zodiacal light (the sunset glow best seen from equatorial latitudes caused by dust in the Solar System reflecting sunlight, not to be confused with the more widely seen sunset glow caused by the Earth's atmosphere reflecting sunlight). Some scientific work has also been carried out by the astronauts on the *Mir* spacecraft, the largest and most recent of the Soviet stations in space, in particular 115 observations of the spectacular supernova in February 1987.

The *Salyut* series has also been used to amass experience of prolonged spaceflight and its

problems. A number of Soviet astronauts have now completed space missions longer than 100 days and just after Christmas 1987 the longest manned mission so far was recorded, involving 326 days in space aboard the *Mir* spacecraft. In general, however, the Soviet space programme is not so heavily committed to the advancement of astronomical knowledge as is the American initiative.

The advantage of having space-based crews working in co-operation with ground-based experts is that more scientific and industrial research will be achieved, and faster, than if the equipment is used by remote control from Earth. The first time the Americans tried this approach was in their Skylab craft in 1973–4. Orbiting 450 kilometres above the Earth, it was both 'office' and 'home' for three teams of astronauts. Each mission was longer than the one before, increasing from 28 to 59 to 84 days. Together the teams orbited the Earth 2476 times and spent over 3000 hours conducting eight categories of experiments. The information they returned included nearly 300,000 pictures of the Sun and 46,146 frames of Earth. On the third mission the crew extensively observed and photographed Comet Kohoutek.

Skylab was launched in the traditional way by using Saturn rockets, a very costly exercise in that very little of the rocket is reusable. In contrast, although the Space Shuttle takes off like a rocket, it travels around the Earth like a spaceship and lands like an aeroplane. Unlike their predecessors, which were only good for one flight, the Shuttle launchers can be used again and again. The orbiter which carries the crew and the cargo is designed to last for a minimum of a hundred flights and the rocket-boosters, parachute and other equipment for around twenty

The American Skylab craft operating in the early 1970s offered one of the first opportunities for astronaut-scientists to work in space.

launches. At present the Shuttle flights last from two to ten days, but it is hoped they will be extended to around thirty days. Moreover, because of the relatively low acceleration and deceleration forces exerted on crew and passengers during launch and re-entry, scientists with only a minimum of astronaut training can use this transport system, provided they are in good health. It can also carry much larger crews into orbit than was previously possible and has a wide range of possible applications, including launch-

ing satellites for scientific and commercial use, repairing spacecraft in orbit and returning satellites to Earth for more extensive refurbishment and relaunch. The cargo, whether satellites, instruments, or a complete unit such as Spacelab, is carried in the large (18 × 4.6 metres) pay-load bay immediately behind the crew's cabin.

Spacelab is an improved version of Skylab, with an estimated life of fifty missions. Two complete prototypes have been built by ESA, the second financed by NASA. These small but well-equipped laboratories have been specially designed for use in zero gravity and feature a number of interchangeable elements which are put together to meet the needs of a specific flight.

The Earth's atmosphere masks our view of the Sun's outer corona, revealed here in all its glory in a Skylab image, one of those obtained in a nine-month study of the Sun.

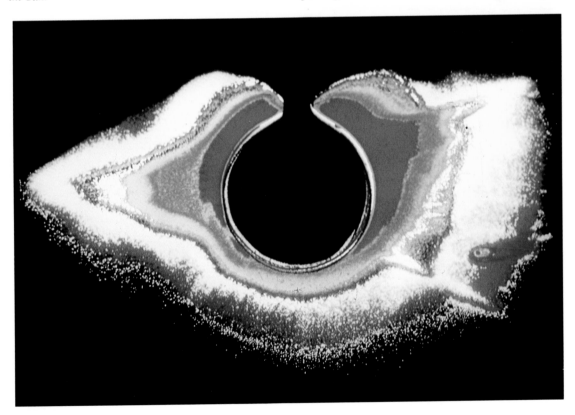

Apart from the laboratory area itself, there are pallets which can be placed in the cargo bay behind the laboratory module to hold the instruments which require direct exposure to space. Spacelab can consist of the laboratory module with or without one or up to five pallets, or of any number of the pallets on their own. Four people can eat and sleep in the orbiter and perform their scientific tasks in the laboratory without having to wear space-suits, using the high vacuum and microgravity of orbital space to make observations free from the Earth's obscuring atmosphere. When the orbiter returns to Earth with its load, the instruments aboard the Spacelab can be removed and changed for the next programme and a new range of experiments. The Spacelab facilities are available to scientists from all over the world, whether they are attached to research institutes and laboratories, government bodies, commercial organizations, or simply have an individual interest in space research. Many of the missions are government sponsored but Spacelab is open to all applicants.

Spacelab 1 carried 38 experiments designed by scientists from 13 nations, of which seven were devoted to astronomical studies. Two of these monitored the Solar Constant (the amount of solar radiation received at the top of the Earth's atmosphere), a third recorded the solar spectrum. The others – involving a far ultra-violet camera, an X-ray spectrometer, a very-wide-field ultra-violet camera, and a cosmic ray detector – all monitored a range of radiation from the stars. The results were to be used to study the structure of our Galaxy and to determine the characteristics of the various forms of radiation, which it was thought would yield clues about how stars evolve. All the experiments were designed to exploit the advantage of observation from orbit, well above the Earth's obscuring atmosphere, but there were problems with some of the hardware and the trip was not as successful as it might have been.

Some scientists were also disappointed with the results from Spacelab 2, which was primarily dedicated to astronomy and included the largest number of astronomers ever assembled together in orbit. From NASA's point of view the mission was a success because the crew was able to point out a number of problems with the Spacelab system which could be rectified for future missions. But it was not a resounding success for the scientists. The pay-load bay of the Shuttle was completely filled with fourteen experiments, while the personnel remained in the orbiter part of the craft operating the instruments by computers. The instrumentation included telescopes for studying the Sun, an X-ray telescope for observing clusters of galaxies and a helium-cooled infra-red telescope for studying interstellar clouds. The pay-load bay was very tightly packed and the crew found it was not possible to operate all the instruments at once as they had to make sure working parts did not collide with one another. One instrument that did work perfectly was the Birmingham University X-ray telescope, which observed the radiation emitted by the extremely hot gas found in clusters of galaxies, as well as observing the centre of our Galaxy. This instrument and the other thirteen were removed from the Spacelab when it returned to Earth. They all have the chance of flying again on future missions, but in the meantime problems will be sorted out and the data received analysed.

Although the idea of an American space station has been around for a number of years, it was formally announced by President Reagan in January 1984 as the, then, next major space objective for the United States. He directed NASA '. . . to develop a permanently manned space station and to do it within a decade'. It is to be a permanent multi-purpose facility in orbit 400 kilometres above the Earth. It will act as a laboratory, an observatory, a garage for fixing and servicing other spacecraft, an assembly plant to build structures too large to be transported in one piece by the Shuttle, and a storage

If scientists are to work in space for weeks or months on end it is essential that they are able to keep fit and healthy. Here Charles Conrad is being examined aboard Skylab by Dr Joseph Kerwin, the weightless conditions making the doctor's task easier than it would be on Earth.

warehouse for satellite parts, amongst other things. There will be manned and unmanned elements. The initial space station will support a crew of eight, who will work there for around three months before the Shuttle flies out a replacement crew.

The system will be designed and built from ready-made sections which will be transported into space in the Shuttle's cargo bay. The work of unloading the parts and assembling them will be done by space-suited astronauts wearing gas-powered backpacks to propel themselves around. The main tool aiding a Manned Manoeuvring Unit (MMU), as the human spacecraft is called, is a long crane like a mechanical arm and hand which can be extended from the Shuttle's cargo bay. This will be used to move floating modules and other components into the required positions so the MMUs can join them together. A similar system will be used once the space station is operational to carry out maintenance work and repairs, or to modify the station by adding or exchanging components. The crane and the MMUs have already been tried and tested during Shuttle flights in the early 1980s and the Shuttle's

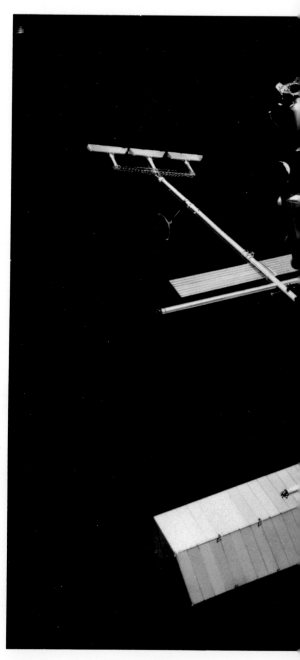

ABOVE *Powered backpacks such as that worn by this astronaut will give those working in space freedom to manoeuvre without being hampered by being tied to their spacecraft.*

RIGHT *The planned American space station will offer scientists the opportunity of living and working in space for weeks or months at a time. Equipment and personnel will be transported to and from the station by the Space Shuttle.*

ability to retrieve and repair satellites has also been demonstrated more than once. In April 1984, for example, an instrument aboard the Solar Maximum Mission satellite (so named because it observed the Sun during one of its 11-year periods of maximum sunspot and flare activity) was repaired whilst in orbit.

Although the final configuration of the first space station is not yet decided, it will certainly have modules rather like aeroplane cabins about 15 metres long by 5 metres wide which will be used as living-quarters and laboratories. Here the crews will be able to work in Earth-like conditions unhampered by space-suits. Attached to these modules will be platforms or pallets housing automated or remote-controlled equipment that needs to be open to space. Other instruments can be carried on unattached, free-flying platforms travelling in their own orbit nearby, which allows observations to be made without any possible interference from the station itself. The crew will be able to move easily along connecting tunnels from one module to the next, and to reach all the external instrumentation by becoming an MMU.

One advantage of having the scientists with their experiments and equipment permanently in orbit is that they can operate programmes which require on-site monitoring and adjustment over long periods. It will no longer be necessary to return an experiment to Earth after a mission in order to prepare it for relaunch. The crew will have the ability to deal with any problems on site or to make immediate changes in their plans in the light of results, particularly as they will be partly independent from ground control. The crew will be largely left to use the equipment as they think fit and to carry out their work when and how they wish. They will be backed up by ample power supplies for their living and research needs, and by modern data processing and communications systems.

Essentially the space station will act as a workshop in space, for on-board experiments and for servicing astronomical satellites. For example, the Hubble Space Telescope could be taken there for overhaul instead of returning it to Earth, thus saving time and money. The station will also provide servicing facilities for missions destined for other planets. The space station is without doubt a very important element in NASA's future plans, a major step – together with the Shuttle – in the completion of its Space Transportation System (STS), which is being designed to give optimum flexibility and opportunities to reuse components. For the first crews using the station, which NASA hopes will be in the 1990s, life will be much like living on an off-shore oil-drilling platform. In the future, working in space may become as familiar as going to offices and workshops on Earth.

8 · Exploring Space

The second half of the twentieth century is often referred to as the space age. It is true that we have gone some way towards conquering space with successful explorations of our local space neighbourhood, but there is plenty more to explore. Just as our ancestors once dreamed of travelling to the Moon, so now we are facing the challenge of looking beyond the Solar System. In almost thirty years we have sent probes to nearly all the planets, to Mercury, Venus, Mars, Jupiter, Saturn and Uranus, and one is now on its way to Neptune. Landers have touched down on Venus and Mars. Orbiters have been sent around the Moon, landers have touched down on it, and men have returned to Earth with rock samples from its surface. Space probes have also been sent to that famous Solar System traveller Halley's Comet. These space missions divide into the manned missions concentrating on the Moon, and the unmanned missions to the planets.

Man and the Moon

Whether you are old enough to remember seeing the first man to set foot on another world, everyone knows that Man has been to the Moon. Many people have now seen some of the Moon rock that was brought back and some can even repeat Neil Armstrong's words as he stepped off the lunar module on to the Moon's surface: 'That's one small step for a man, one giant leap for mankind'. Remarkable as it may seem to those of us who were around at the time, it is now twenty years since that historic moment and although enormous moves forward have been made in space technology, the Moon landing is still regarded as Man's finest achievement to date in the quest to put himself into space. This is partly because in the last two decades the US space programme has concentrated on unmanned missions, rather than on trumping the Moon landing. It was the dream of the 1960s, and of the American nation and its President John F. Kennedy in particular, to land a man on the Moon. As Kennedy's May 1961 speech proclaimed, 'this nation should commit itself to achieving the goal, before the decade is out, of landing a man on the Moon and returning him safely to Earth'. This dream was achieved largely because it *was* a political aim of the American nation, but it was also an extraordinary opportunity for scientists to develop the necessary technology and for astronomers to learn much about their nearest neighbour in space.

When Kennedy declared his Man-on-the-Moon intentions the science of space travel was still very young. It was only twenty days since Alan Shepard, America's first astronaut, had travelled in space and less than four years since the world's first spacecraft, the Soviet *Sputnik 1*, was launched. In this short period the Space Race had got well and truly under way, with America and the Soviet Union competing strongly to see who could achieve what first. The Soviets had undoubtedly leapt ahead at the starting gun with the launch of *Sputnik 1*, and they were also the first to get a man into space. Major Yuri Gagarin was launched in *Vostok 1* in April 1961, completing a single orbit of the Earth at a maximum altitude of 326 kilometres before returning safely after 108 minutes. His historic flight was a challenge to American expertise and the following month Alan Shepard made America's first manned spaceflight.

Shepard's flight grew into Project Mercury, a programme of manned flights designed to test systems and train pilots for more ambitious future missions. This was followed by the Gemini programme initiated in 1965, which included rendezvous and docking tests and several space walks. Space experience learnt through the American Mercury and Gemini programmes and the Soviet Voskhod and Vostok programmes was invaluable for the future. Both nations now pursued more ambitious plans, but while the Soviets concentrated on complex missions orbiting the Earth, the Americans chose to send the first man to the Moon.

The first stage in the Moon programme was to learn more about our nearest neighbour in space. Close-up photographs of the Moon's surface, including the first pictures of the far side of the Moon which is always turned away from Earth, had already been obtained from the Soviet *Luna* probes. The American probes in the *Pioneer* and *Ranger* series were initially less successful, but *Rangers 7, 8* and *9* worked perfectly and together sent back some 17,000 pictures of the lunar landscape. The Americans progressed from this photographic study to sampling the Moon itself through their *Surveyor* and *Orbiter* probes. Five of the seven *Surveyor* craft soft-landed on the Moon between 1966 and 1968, equipped with movable television cameras and soil samplers and able to measure surface radiation, soil structure and local magnetic fields. While the *Surveyor* missions proved that it was feasible to land astronauts on the Moon, the five *Orbiters* launched between August 1966 and August 1967 photographed a

From the first historic landing on the Moon in July 1969 to the last manned mission in December 1972, American astronauts travelled across almost 90 km of the Moon's surface in not quite 80 hours.

number of potential landing sites for the astronauts, as well as imaging 99 per cent of the Moon's surface and providing detailed coverage of 36 sites of scientific interest. Meanwhile the Soviets were showing their expertise in landing probes, collecting lunar samples and returning them to Earth. Their robot Moon-car Lunokhod 1, operated from Earth by remote control, travelled over 10 kilometres and mapped 79,000 square metres of the lunar surface in 1970 and 1971.

The *Apollo* series of spacecraft was chosen to take Man to the Moon. In December 1968 the astronauts Frank Borman, James Lovell and William Anders were launched in *Apollo 8*, the first time that anyone had left the vicinity of the Earth. In the following year came *Apollo 11* and the lunar landing mission commanded by Neil Armstrong. While Michael Collins stayed in the command module orbiting the Moon, Armstrong

This lunar boulder, about 1.5 metres long, was found in February 1971 by the Apollo 14 *astronauts.*

and Edwin 'Buzz' Aldrin journeyed to the lunar surface in the lunar module, codenamed *Eagle*. In late July 1969 President Kennedy's dream came true, Man was on the Moon. Although other *Apollo* landings followed, no subsequent mission caught the attention of the world in quite the same way. Man had met the challenge of transporting himself to another world.

But, despite the razzmatazz, the *Apollo* programme was also concerned with collecting scientific information and the *Apollo 11* crew set up a number of experiments on the lunar surface. A reflector was positioned pointing towards the Earth so that a laser beam could be bounced off it and give a precise measurement of the Earth-Moon distance. A seismometer was set up to measure the frequency and level of earthquake-like shocks on the Moon, while other instrumentation investigated the characteristics of the solar wind (see *The Greenwich Guide to the Planets*), from which we are shielded by the Earth's atmosphere.

The 22 kilogrammes or so of lunar rock samples that Armstrong and Aldrin collected were of particular interest. This rock and the other 2000 or so samples brought back in subsequent *Apollo* missions, together totalling almost 400 kilogrammes, have given us invaluable information on the composition of the Moon and its history. For example, the *Apollo 11* samples taken from the Mare Tranquillitatis landing site revealed a dust-type rock 5 metres deep. Apart from small craters which had marked the surface, this material appeared to have been hardly disturbed in the past hundred million years. This mission also obtained fragments of basaltic lava which had cooled and crystallized

RIGHT *Almost 400 kilogrammes of lunar rock have been brought back to Earth by* Apollo *astronauts. This field of boulders on the flank of Cone Crater was surveyed by Alan Shepard and Edgar Mitchell, two of the* Apollo 14 *crew.*

3700 million years ago, a sample of the flows which form the features known as *maria* (see *The Greenwich Guide to the Planets*).

Among later trips to other landing sites, one of the most informative was the *Apollo 17* mission, which landed in the mountainous highlands bordering the Mare Serenitatis. The 120 kilogrammes of moon rock brought back to Earth by this spacecraft was the most varied collection of any of the *Apollo* missions, the range of samples indicating the main stages of the Moon's history since it was formed 4600 million years ago. The mixture included breccias formed of cemented rock fragments, basaltic lavas from the time when the Moon was still young and its interior molten, and surface rock which had been produced as a result of meteorite bombardment over the past 3800 million years (see *The Greenwich Guide to the Planets* for a full account of the Moon's evolution).

Planetary probes

Our present knowledge of Solar System planets other than the Earth owes much to the interplanetary space probes that have visited them in the last 25 years. As information has been obtained and objectives have changed, the probes themselves have grown in sophistication and complexity. Not surprisingly the first targets for investigation were the two planets closest to Earth, Venus and Mars. Despite its proximity and the fact it is a similar size to Earth, Venus was, and still is to a certain extent, a mystery planet, as the surface is thickly shrouded in impenetrable cloud. It has proved to be one of the most difficult planets to study, whether from Earth or space.

The American *Mariner 2* spacecraft, which reached Venus in August 1962, revolutionized our knowledge of the planet and put paid to any thought of Venus being like Earth, despite its similar mass and chemical composition; surface temperatures of at least 425°C were indicated, hot enough to melt lead. Useful results were also obtained from the Soviet *Venera* series of spacecraft, the first of which dropped instrument packages into the Venusian atmosphere to investigate its composition. Scientists had expected the atmosphere to be like the Earth's, largely made up of nitrogen with a little carbon dioxide, but in 1967 *Venera 4* found that it was in fact well over 90 per cent carbon dioxide, with small quantities of oxygen and water vapour but no nitrogen. Then came *Venera 9*, which succeeded in landing instruments on Venus' surface on 22 October 1975, sending back what was then our only photograph of the planet's barren and rocky terrain. Only three days later *Venera 10* sent back the second surface picture, showing an older landscape than that surveyed by *Venera 9*, a weathered plateau with flat slabs of rock. The first colour images came in 1981, and then in 1983 *Veneras 15* and *16* radar-mapped Venus' surface, revealing impact craters, hills, ridges and other major topographical features.

The first details of the planet Mercury were obtained in 1974 when the spaceprobe Mariner 10 *flew within 703 kilometres of the planet.*

One of the Venus probes, *Mariner 10*, also gave us our first close-up view of Mercury, the planet closest to the Sun, completing three fly-bys during a one year period in 1974–5. The first images were taken from a distance of 5.3 million kilometres and were no better than Earth-based views, but as the craft moved in towards the planet more detail was revealed and Mercury appeared increasingly Moon-like. Vague white spots resolved into craters, ridges and lava-flooded areas became clear, and a huge meteor-like impact basin was revealed, now known as the Caloris Basin. A better understanding of Mercury's composition was reached when it was realized that its density is almost the same as Earth's and that it too must have a metallic core. In other respects, however, Mercury is very different from our own world. The night-time temperature was recorded as −183°C and the maximum daytime temperature as 187°C. Two further fly-bys produced more images of the planet, but although about half the surface has now been covered Mercury may still have surprises in store for us.

In comparison Mars has been well investigated by both American and Soviet probes. The American *Mariner* series provided us with our first close-ups of this planet, pictures that were to dispel any expectation that life could exist there. *Mariner 4* was the first successful Mars fly-by, approaching to a distance of 9600 kilometres in July 1965. The pictures it sent back – 21 partially overlapping shots, together making up 1 per cent of the surface – showed heavily cratered lunar-like regions, chaotic terrain and desert-like flat areas, all suggesting that Mars was a dead world. *Mariners 6* and *7* went even closer, within 3500 kilometres, *Mariner 6's* more detailed pictures largely confirming what *Mariner 4* had already shown. But they also included images of the

south polar cap which prompted scientists to reprogramme *Mariner 7* to obtain more pictures of this area. In November 1971 came *Mariner 9*, which changed views of Mars once more. A photographic survey of the complete surface of the planet carried out by this probe revealed hitherto unknown types of terrain, including four huge volcanoes, the largest of which, now known as Olympus Mons, towers some 25 kilometres high and has a diameter of 600 kilometres. To the south-east of the volcanoes was another specta-

cular feature, the Valles Marineris rift valley, 4000 kilometres long, 100 kilometres wide and 6 kilometres deep in places. These features convinced scientists that Mars is a geologically active planet, not the moribund world the first pictures had suggested.

The next step was to land instruments on Mars itself. The first soft landings were achieved by the Soviets as early as 1971, but only 20 seconds of television signals were recorded before transmissions were suddenly discontinued. The fol-

During the summer of 1976 two Viking *craft landed on Mars and sent back the first television images of the surface; instruments also analysed the content of the Martian rock.*

This composite picture of the 272-kilometre Mangala Valles on Mars − the deep valley just below the two central craters − was built up from a series of Viking *images.*

lowing four Mars craft launched by the Soviets in 1973 were equally beset with problems, and it was another three years before two American *Viking* landers touched down on Mars in the summer of 1976, their landing sites having been carefully chosen with the aid of the *Mariner 9* photographic survey. These two landers have proved the most successful to date in relaying information about Mars. On its eighth day on the surface *Viking 1* used its 3-metre robot arm to collect the first rock samples, which it was hoped would reveal evidence of the existence of life. Aluminium, silicon, calcium, iron, carbon dioxide and water were all detected when the samples were analysed on the spot, but no organic material, to the great disappointment of the biologists. Other experiments in the search for life were carried out, but no firm conclusions could be drawn. Together the two *Viking* landers returned 4500 pictures, the last from *Viking 1* in November 1982. Their parent orbiters produced over 51,000 images of the planet's surface.

The first craft to voyage further out into the Solar System were the American *Pioneers 10* and *11*, launched in 1972 and 1973, whose journeys to Jupiter and Saturn took them further than any man-made object had ever been. *Pioneer 10's* initial speed was 52,000 kilometres per hour and *Pioneer 11* travelled at an incredible 172,000 kilometres per hour. Their journeys were more hazardous than those of earlier craft as they had to survive the rock- and dust-filled asteroid belt where it was feared a particle as small as 0.05 centimetres could end the mission if it collided with spacecraft travelling at such enormous speeds. There was also the unpredictable effect of Jupiter's radiation to contend with. If all went well the *Voyager* spacecraft were to follow in a few years time to carry out a more thorough survey.

It took *Pioneer 10* almost eight months − from July 1972 to February 1973 − to journey through the 280 million kilometres of the asteroid belt, but fortunately the dust hazard was found to be not as serious as expected and special cells mounted on the underside of the craft detected only 55 impacts with dust particles. Both *Pioneers* have instruments on board to supply information on magnetic fields, radiation and dust in interplanetary space. *Pioneer 10* should continue to send back signals until 1994, when it will be in interstellar space and moving away from Earth in the direction of the constellation Taurus at the rate of 360 million kilometres a year. *Pioneer 11* will cross Neptune's orbit in 1990 and head towards the centre of our Galaxy in the direction of the constellation Sagittarius (i.e. in the opposite direction to *Pioneer 10*). In 1991 it too will be in interstellar space and it is expected to keep transmitting until about 1994. Should either craft ever be retrieved by intelligent life, each of

them carries a gold-coated aluminium plaque with diagrams explaining where the craft came from and showing male and female humans.

Images of Jupiter were obtained by both *Pioneer* craft. In fact *Pioneer 10*'s flight was so successful that *Pioneer 11* was reprogrammed to take a much closer view of the planet before moving on to give the first close-ups of Saturn. There were

The great gas giant Jupiter, visited by Pioneers 10 *and* 11 *and* Voyagers 1 *and* 2, *has proved to be one of the most picturesque of the planets. This image of Jupiter's satellite Io seen against the planet's turbulent clouds was captured by* Voyager 2 *on 25 June 1979 from a distance of 12 million kilometres.*

fears that it would receive a severe battering when it encountered Saturn's rings, but, as in the case of the asteroid belt, these fears proved unfounded. What *Pioneer 11* did see was another narrow band just outside the main Saturn ring system, and a new satellite was also spotted, but many, many more details of Saturn, its rings and its satellites were to be revealed by the more thorough *Voyager* missions.

In the late 1970s the planets were in a pattern which only occurs about every 175 years and allows a spacecraft launched from Earth to Jupiter to also visit Saturn, Uranus and Neptune, gathering speed as it passes each planet. Scientists took advantage of this infrequent planetary alignment to launch the two *Voyager* missions in 1977. The identical craft were designed to withstand the rigours of extended travel in outer space, to deliver scientific information back to Earth and also to respond to commands from Earth, although an on-board system of sensors and computers also enabled

This dramatic volcanic eruption on Io, recorded by Voyager 1 on 4 March 1979 from a distance of about 500,000 kilometres, is an example of the spectacular revelations about Jupiter's satellites obtained from the Voyager spacecraft. By studying colour images like this astronomers can estimate the amount of gas and dust in the volcanic explosion.

them to care for themselves. Each craft works on a power supply of 400 watts provided by nuclear generators, using only 25 watts – much less than that required by a household lamp – to transmit messages to Earth.

Voyager 1 was launched second because it was designed to follow a faster route, overtaking *Voyager 2* in the asteroid belt. It took its first views of Jupiter from a distance of 265 million kilometres and in early 1979, as it moved in for its closest fly-by on 5 March, it transmitted pictures every two hours to produce time-lapse studies of the planet's rotating, swirling atmosphere. It also took close-ups of Jupiter's four Galilean moons, Io, Europa, Ganymede and Callisto. *Voyager 1* then moved on to Saturn, making its closest approach on 12 November 1980. It showed that the ring system consists of thousands of concentric bands made up of particles ranging from 10 metres to 0.0005 centimetres in size.

Saturn imaged by Voyager 1 *on 13 November 1980. The planet was deliberately overexposed in order to bring out detail in the rings.*

Other revelations included the discovery of three additional Saturnian moons, two Jovian moons and, most interestingly, that Jupiter, like Saturn, is a ringed planet.

Voyager 2 added another satellite to the Jupiter list and also showed how both Jupiter and Io had changed in four months since the earlier fly-by. The Great Red Spot on Jupiter had become more uniform in appearance and some of the eight volcanoes that had been erupting on Io – the first active volcanoes ever seen beyond Earth – were now dormant. *Voyager 2* reached Saturn in 1981

Craters 2.5 kilometres in diameter are revealed in this Voyager 1 *image of Saturn's moon Rhea taken on 12 November 1980. This is one of the most heavily cratered parts of Rhea's surface.*

and in the light of information from *Voyager 1* it was used to investigate the planet's ring system thoroughly. Profiting from their earlier experience the scientists were able to set camera exposures at more exact levels and so obtained finer detail than had been received from *Voyager 1*. By observing the star Delta Scorpii through the rings scientists were able to get a much more accurate picture of the ring system, as the flickering starlight identified every tiny ringlet. There is still plenty of data here to process and analyse in the years ahead. Altogether the *Voyagers* took some 33,000 photographs of Jupiter and its satellites and 30,000 of the Saturnian system, study of which has already revealed more Saturnian satellites, giving a current confirmed grand total of 21.

In January 1986, after a journey of 5000 million kilometres, *Voyager 2* reached Uranus, the mysterious planet discovered in 1781. The probe revealed ten new Uranian moons and details of its rings, but the planet's thick atmosphere obscured any view of the surface. The signals from Uranus took $2\frac{1}{2}$ hours to cover the 2880 million kilometres to Earth. *Voyager 2* is now moving further away from Earth towards Neptune and one of its satellites, Triton, and it will then join *Voyager 1* and the two *Pioneer* craft in investigating the outer heliosphere, the region where the solar wind starts to intermingle with the material of interstellar space. *Voyager 1*'s messages should be powerful enough to reach us until 2012, when it will be at a distance of 18,000 million kilometres, and *Voyager 2*'s for a further year, fading at a distance of 16,000 million kilometres. Like the *Pioneer* craft, each *Voyager* carries a message from Earth. It is inscribed on a gold-plated copper disc like a gramophone record. If anyone ever uses the enclosed needle to listen to it, they will hear 90 minutes of music, 115 analogue pictures, greetings in 60 languages and other natural Earth sounds. The record has an aluminium cover which will keep it safe for 100 million years.

The most spectacular results from spacecraft missions since the *Voyager* encounters were those obtained from the probes to Halley's Comet in 1986. Of all these, including the Soviet probes *Vega 1* and *2*, and the Japanese *Sakigake* and *Suisei*, the most praiseworthy was the European *Giotto*, whose success firmly established the European Space Agency as a force in modern space exploration. *Giotto* travelled considerably closer to the comet's centre or nucleus than any of the other craft. The images sent back to Earth as it flew into its target on 13 March were at first difficult to interpret, as the camera automatically followed the brightest object in its field of view which made it difficult to see the comet's dark nucleus. But after processing the information received in the control room at Darmstadt, West

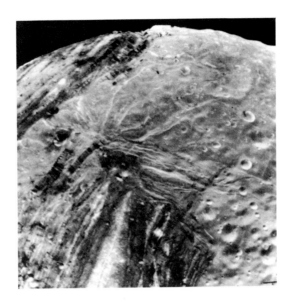

Without the information obtained from the Voyager 2 *spacecraft, our knowledge of Uranus and its satellites would be much the same as what was known 150 years ago.* Voyager 2*'s image of Miranda, one of Uranus's larger satellites, reveals varied geological structures pointing to a complex history.*

Germany, a cometary nucleus was seen for the very first time.

The heart of Halley's Comet proved to be an irregular-shaped body like a peanut, measuring about 15 kilometres long by 9 kilometres wide. It was very similar to the dirty snowball that had been expected (see *The Greenwich Guide to the Planets*). As *Giotto* flew closer and closer to the nucleus, within the comet's coma, dust particles collided with the spacecraft at an ever-increasing rate. Just 14 seconds before the scheduled closest approach *Giotto* was struck by a dust particle big enough to interrupt the mission. Half an hour later control of the craft had been regained but no more pictures were returned. On 15 March *Giotto*'s experiments were turned off and it is presently hibernating in orbit around the Sun.

The future

There are firm plans for Man's future exploration of space and some of the instrumentation and craft which form part of these plans, such as the Hubble Space Telescope, have already been built. A number of planetary missions are in the sights of the Soviets, the Americans, the Europeans and the Japanese. NASA's Magellan craft which will radar-map Venus at high resolution is expected to be launched soon and the Japanese are looking to the 1990s to launch their *ISAS* Venus craft. Mars is once again going to be the subject of some intense investigation. Two Soviet *Phobos* probes were launched in July 1988 to journey to Mars to study the moon after which they are named and possibly also Deimos. A further Soviet probe, *Vesta*, will be launched in the 1990s to deploy balloons and penetrators to Mars. The same probe may also investigate comets and asteroids. The Soviets also intend to carry out geochemical mapping of the entire Moon by orbiter in the 1990s. The Americans have similarly ambitious plans, such as the *Galileo* mission to Jupiter taking in an asteroid on the way, but these have been delayed by the problems with the Shuttle. ESA's *Ulysses* solar probe, which was to fly high above the ecliptic plane and the Sun's poles for the first time, was to be launched at the same time, but it, too, is in storage.

According to the latest plan the *Galileo* craft will first travel towards the inner Solar System. It will use some of its instruments to study Venus as it moves past this planet. After looping around the Sun it will pass Earth again using the Earth's gravity to increase its speed. Then, after voyaging among the asteroids for nearly a year, *Galileo* will swing by Earth once more, again increasing its speed, and head towards its goal, Jupiter. Once close to the planet a probe will be jettisoned to drop through its atmosphere and the craft will settle into orbit. For just under two years *Galileo* will study the parent planet and its moons and will provide us with much information to add to

The Americans intend to return to the Moon and to establish a permanent base there by the early years of the twenty-first century. In this artist's impression a lunar surface crane removes a newly-arrived habitation module from a lunar lander craft.

our knowledge of this gas giant and its family.

At the beginning of 1988 the Americans announced their distant-future plans. These are based on the initiatives put forward by the former *Shuttle* astronaut Sally Ride, who headed a team investigating the future of the American space programme. Four programmes were outlined; a global study of Earth, an advanced programme of Solar System exploration, a permanent return to the Moon, and a pioneering manned journey to Mars.

The Solar System exploration is to be in three main areas. The Comet Rendezvous Asteroid Fly-by (CRAF) mission aims to investigate the beginnings of our Solar System by studying an asteroid and a comet, both bodies which are thought to be made of material dating from the birth of the Solar System. The second area of exploration is a project to reach Saturn in the early twenty-first century and to carry out a three-year study of the planet and its rings, using an orbital spacecraft named *Cassini* and three probes. One of these probes will go to Saturn's largest moon, Titan. The final mission is the Mars Rover/Sample Return programme, made up of three separate Mars missions all scheduled for around the turn of the century. A robotic surface-rover is planned to collect samples from the planet, which could mean that we shall have a handful of Mars back on Earth by the year 2000.

NASA's 'permanent return to the Moon' programme should also have Man back on the Moon by the year 2000. Once there, astronauts will construct an outpost where they can stay for one or two weeks carrying out scientific investigations. In between times they will be transported to and from the space station orbiting Earth, but eventually it is hoped to set up a

permanently occupied base on the Moon. The experience, expertise and confidence that this programme would bring could then be used in NASA's most exciting and bold project, that of landing men and establishing an outpost on Mars in the twenty-first century. Robotic exploration of the planet and a study of long-duration spaceflight would come first, followed by piloted one year round-trip missions to Mars and the eventual setting up of an outpost on Mars. These missions will not only provide astronomers with invaluable information about our nearest neighbour in space, but inspire people the world over just as the Moon missions of the 1970s did.

Getting Involved

If you want to become actively involved in astronomy you could start by contacting an astronomical society. There are three national bodies in the UK, the Royal Astronomical Society, the British Astronomical Association (both of which have already been mentioned in Chapter 1) and the Junior Astronomical Society, whose membership includes people of all ages despite its name. The national societies tend to be London-centred and to meet infrequently in the provinces, but all three have publications which make the membership fee worthwhile. In addition there are numerous local societies, which usually meet once a week. Some have their own newsletters or journals, and some also have a telescope which can be used by society members. The local library will often have information on the society in your area, but if not the Federation of Astronomical Societies should be able to help you. Although this information is specifically relevant to residents of the United Kingdom, a similar pattern of organizations can be found in countries as far apart as Japan, France and the United States.

In addition, there are a number of observatories, museums, planetaria and space centres around the world which can be visited. A number of the older observatories have now been converted into museums. Although these are generally presented as places where you look and learn rather than actively get involved in astronomy, many have open days or nights when it is possible to take part in an observation programme. The museum at the Old Royal Observatory at Greenwich, for example, tells the story of astronomy at Greenwich from 1675, the date the observatory was founded. The displays include an introduction to the 28-inch telescope, the largest refracting telescope in the United Kingdom and the seventh largest in the world. Although this telescope is a museum exhibit it is still very much a working instrument. It is available for societies to use and it is also used for the open evenings held at Greenwich during the winter. The observing programme on these evenings may include close views of planetary bodies, such as Jupiter and the Moon, as well as looking further afield to nebulae and galaxies. When there is something unusual to be seen, such as a comet, additional observing evenings are arranged.

It is also possible to visit a number of working observatories, such as the Kitt Peak National Observatory in Arizona. As scientists are at work here all the time, not all the observatory is open, but you can watch instruments being used from special galleries and on occasional evenings throughout the year visitors are offered the chance to take part in the observations. Should you ever find yourself in Hawaii, you could visit the Mauna Kea Observatory where there are both optical and infra-red telescopes. Be warned, however, that the observatory is sited on top of a 4200-metre extinct volcano, high enough to give you headaches and nausea. Also you have to find your own way of getting up there. The UK's Jodrell Bank in Cheshire, an internationally-known centre of radio astronomy, is rather more accessible.

Many people find visiting planetaria which

simulate the night sky as exciting as going to a working observatory. Images of the stars and planets are thrown on to a domed surface above your head, giving you an unparalleled view of the heavens. The planetarium's great advantage is that it can be set to show the sky from any position on Earth for any date and time. A specialized cinema screen that simulates what it would be like to travel in space is a recent development of the planetarium concept. There are now a number of these Omnimax and Imax theatres in the United States, and others at The Hague in Holland, the Science Centre at La Villette in Paris and in Hong Kong and Singapore.

To see hardware that has actually been into space, or to get a close look at some Moon rock, it is best to visit a specialized museum like the National Air and Space Museum in Washington, or the space gallery of the Science Museum in London, or better still one of the space centres in the United States. At the John F. Kennedy Space Center in Florida you can watch huge Imax cinema displays, and tour the various rocket pads which launched Man to the Moon and are now a vital part of the Space Shuttle programme. At another working space centre, the Lyndon B. Johnson Space Center in Houston, Texas, visitors can see the Mission Control Center which controls all American manned space flights and peek into the laboratories where many of the Moon rock samples are under investigation. Information about the Soviet space programme is on display at the Space Museum and Park in Moscow.

If after visiting observatories and museums and joining your local club, you want to *be* an astronomer, what should you do? As most professionals are specialists in one area of astronomy, applying skills in mathematics and physics to their particular specialization, it is best to concentrate on gaining good qualifications in these two subjects rather than learning the names of the stars, or trying to remember the distance of the Andromeda Galaxy and how long it takes Jupiter to revolve around the Sun. With them you can apply for a place at a university to take a degree in astronomy, physics or mathematics, or a combination of these three disciplines. An aspiring professional astronomer would then need to take a further degree, involving original research. Only then would he or she be ready for their first post.

If this route is not for you or seems much too daunting, remember that astronomers could not function without their support teams of engineers, computing experts, photographers, journalists, secretaries, and other specialists. There are also endless opportunities to contribute as an amateur. If you really want to become involved in the astronomical world, you will find a niche somewhere. As a start, you could read all four guides in this series, as together they give you a complete introduction to the night sky and the world of astronomy.

Some useful UK addresses

British Astronomical Association,
Burlington House, Piccadilly,
London W1V 0NL

Federation of Astronomical Societies,
Whitehaven,
Maytree Road,
Lower Moor,
Pershore. Worcs WR1D 2NY

Junior Astronomical Society,
36 Fairway,
Keyworth,
Nottingham NG12 5DU

Royal Astronomical Society,
Burlington House, Piccadilly,
London W1V 0NL

Index